John Ponsonby Cundill

A Dictionary of Explosives

John Ponsonby Cundill

A Dictionary of Explosives

ISBN/EAN: 9783337337612

Printed in Europe, USA, Canada, Australia, Japan

Cover: Foto ©berggeist007 / pixelio.de

More available books at **www.hansebooks.com**

A

DICTIONARY OF EXPLOSIVES

(SECOND EDITION)

BY

LIEUT.-COL. J. P. CUNDILL, R.A.,

LATE

ONE OF HER MAJESTY'S INSPECTORS OF EXPLOSIVES.

ENTIRELY RE-ARRANGED AND BROUGHT UP TO DATE

BY

CAPTAIN J. H. THOMSON, R.A.,

H.M. INSPECTOR OF EXPLOSIVES.

LONDON:
PRINTED FOR HER MAJESTY'S STATIONERY OFFICE;
AND SOLD BY
EYRE AND SPOTTISWOODE,
EAST HARDING STREET, FLEET STREET, E.C.

1895.

Price Twenty-one Shillings.

CONTENTS.

PREFACE TO SECOND EDITION.

It is much to be regretted that Colonel Cundill was prevented by his long illness and untimely death from bringing out this edition of his valuable work. It had been his intention to re-arrange the Dictionary in alphabetical order, the subdivision into classes having been found to be inconvenient. This has accordingly been done. The book has been brought up to date (December 1894) and some additions have been made where it was thought they would be useful. In this I have been much assisted by the Translation which M. Desortiaux, of the French War Office, kindly placed at my disposal, and in which the work had been so re-arranged and brought up to date.

J. H. THOMSON,
Capt. R.A.,
H.M. Inspector of Explosives.

Home Office,
January 1st, 1895.

AUTHOR'S PREFACE.

IN the following pages I propose to set forth a " Dictionary of Explosives," which comprises all those which have, to my knowledge, been used or proposed to be used. The list includes a good many which are obviously unfit for practical purposes, and some which are merely curiosities in a chemical sense. All, however, are given without distinction.

Those which are, or have been, licensed for use in the United Kingdom will be specially noted as "authorised explosives," a term which signifies that they are, or have been,* " authorised to " be manufactured for general sale or imported for general sale," under the Explosives Act, 1875.

It may not be amiss to point out here that the authorisation of an explosive in no way concerns its commercial or practical value. It simply means that a given explosive has, after due examination and testing, been found to be reasonably safe under ordinary conditions of transport, storage, and use.

It will be apparent as the lists are gone through how often the same, or practically the same, idea has been patented and re-patented, and this is especially the case in the chlorate mixture class.

With gunpowder I have dealt very slightly indeed, for the literature on that head is very copious, and I have merely indicated a few propositions which have been made with respect to alterations from the ordinary methods of manufacture.

I would call particular attention to the fact that I do not pretend to enumerate every form of matter possessed of explosive properties. Of these the quantity is innumerable, and the

---

* The letter A is also placed against some explosives which have been approved for license but for which no license has actually been granted.

practical chemist is acquainted with scores of combinations which, under certain circumstances, are powerfully and dangerously explosive. To take the simplest instance, a mixture of oxygen and hydrogen in the proportions requisite to form water is, weight for weight, one of the most powerful explosives known; while the very water created by the combination of these gases will, if closely confined and heated, give rise to explosion, as in the case of boiler accidents. Many other well-known gaseous mixtures, such as coal gas and air, are highly explosive, but such will not be mentioned here. Some of the more interesting chemical combinations are given, but, as a rule, I have followed the definition given in the Explosives Act, 1875, sec. 3, which states:—

"The term 'explosive' in this Act,—

(1) "Means gunpowder, nitro-glycerine, dynamite, gun-cotton, blasting powders, fulminate of mercury, and other metals, coloured fires, and every other substance, whether similar to those above mentioned or not, *used or manufactured with a view to produce a practical effect by explosion, or a pyrotechnic effect*,* and

(2) "Includes fog-signals, fireworks, fuzes, rockets, percussion caps, detonators, cartridges, ammunition of all descriptions, and every adaptation or preparation of an explosive as above defined."

---

* The point of the definition, in regard to substances not detailed by name, lies in the words which I have placed in italics.

By sec. 104 of the Act power is reserved to Her Majesty in Council to declare that any substance which appears to Her "to be specially dangerous to life or "property, by reason either of its explosive properties, or of any process in the "manufacture thereof being liable to explosion," may "be deemed to be an explosive within the meaning of this Act."

ABBREVIATIONS.

"D. 612."—Traité sur la poudre, les corps explosivs et la pyrotechnie, by Desortiaux (after the work of Drs. J. Upmann and E. Meyer), Dunod, Paris, 1878: page 612.

"T. 106."—Tunnelling, Explosive Compounds, and Rock Drills, by Henry S. Drinker, 2nd Edition, John Wiley & Sons, New York, 1882: page 106.

"M. XIII. 246."—Notes on the Literature of Explosives, by Professor C. E. Munroe, of the U.S. Naval Institute, Annapolis: No. XIII., page 246.

"E. 124."—The Modern High Explosive, by Manuel Eissler. John Wiley & Sons, New York, 1884: page 124.

"B., I. 288."—Sur la force des Matières Explosives, by Berthelot, 3me Edition. Gauthier-Villars, Paris, 1883: Vol. I., page 288.

"O. G."—Mr. Oscar Guttmann.

"H. M."—Mr. Henry Mosenthal.

"P. & S. 517."—French Translation of Cundill's Dictionary brought up to date by M. Desortiaux. Gauthier-Villars, Paris, 1893: No. 517.

"Spec. 1234, 31.8.94."—English Patent Sp cification No. 1234, dated August 31st, 1894.

"Fr. Spec. 123,456, 31.8.94."—French Patent Specification No. 123,456, dated August 31st, 1894.

"**A.** III $_2$."—The **A** signifies that an explosive has passed the Home Office tests, and has been placed on the Authorised List, or may be so placed when licensed. The numerals following signify the class and division (if any) to which it belongs, *e.g.*, Class III., Division 2. (*See* Classification, p. vii.)

"**R.** IV." signifies that an explosive has been submitted for license and rejected, and that it belongs to Class IV.

CLASSIFICATION OF EXPLOSIVES.

The following is the text of an Order in Council (No. 1) made under the Explosives Act, 1875,* classifying explosives :—

For the purposes of the said Act explosives shall be divided into seven classes as follows :—

Class 1 - - -	Gunpowder.
Class 2 - - -	Nitrate mixture.
Class 3 - - -	Nitro-compound.
Class 4 - - -	Chlorate mixture.
Class 5 - - -	Fulminate.
Class 6 - - -	Ammunition.
Class 7 - - -	Firework.

And when an explosive falls within the description of more than one class it shall be deemed to belong exclusively to the latest of the classes within the description of which it falls.

CLASS 1.—*Gunpowder Class.*

The term "gunpowder" means exclusively gunpowder ordinarily so called.

CLASS 2.—*Nitrate-mixture Class.*

The term "nitrate mixture" means any preparation, other than gunpowder ordinarily so called, formed by the mechanical mixture of a nitrate with any form of carbon or with any carbonaceous substance not possessed of explosive properties, whether sulphur be or be not added to such preparation, and whether such preparation be or be not mechanically mixed with any other non-explosive substance.

The nitrate-mixture class comprises such explosives as—

Pyrolithe,
Pudrolithe,
Poudre Saxifragine,

and any preparation coming within the above definition.

CLASS 3.—*Nitro-compound Class.*

The term "nitro-compound" means any chemical compound possessed of explosive properties, or capable of combining with

---

* The Acts of Parliament in force at the present time with regard to explosives are the following :—

The Explosives Act, 1875, regulating the manufacture, storage, and transport of explosives, and Orders made under that Act.

The Explosives Substances Act, 1883, relating to the unlawful use of explosives.

metals to form an explosive compound, which is produced by the chemical action of nitric acid (whether mixed or not with sulphuric acid) or of a nitrate mixed with sulphuric acid upon any carbonaceous substance, whether such compound is mechanically mixed with other substances or not.

The nitro-compound class has two divisions.

Division 1 comprises such explosives as—

Nitro-glycerine,
Dynamite,
Lithofracteur,
Dualine,
Glyoxiline,
Methylic nitrate,

and any chemical compound or mechanically mixed preparation which consists either wholly or partly of nitro-glycerine or of some other liquid nitro-compound.

Division 2 comprises such explosives as—

Gun-cotton, ordinarily so-called,
Gun-paper,
Xyloidine,
Gun-sawdust,
Nitrated gun-cotton,
Cotton gunpowder,
Schultz's powder,
Nitro-mannite,
Picrates,
Picric powder,

and any nitro-compound as before defined, which is not comprised in the first division.

Class 4.—*Chlorate-mixture Class.*

The term "chlorate mixture" means any explosive containing a chlorate.

The chlorate-mixture class has two divisions.

Division 1 comprises such explosives as—

Horsley's blasting powder,
Brain's blasting powder,

and any chlorate preparation which consists partly of nitro-glycerine or of some other liquid nitro-compound.

Division 2 comprises such explosives as—

Horsley's original blasting powder,
Erhardt's powder,
Reveley's powder,
Hochstadter's blasting charges,
Reichen's blasting charges,
Teutonite,
Chlorated gun-cotton,

and any chlorate-mixture as before defined, which is not comprised in the first division.

Class 5.—*Fulminate Class.*

The term "fulminate" means any chemical compound or mechanical mixture, whether included in the foregoing classes or not, which, from its great susceptibility to detonation, is suitable for employment in percussion caps or any other appliances for developing detonation, or which, from its extreme sensibility to explosion, and from its great instability (that is to say, readiness to undergo decomposition from very slight exciting causes), is especially dangerous.

This class consists of two divisions.

Division 1 comprises such compounds as the fulminates of silver and of mercury, and preparations of these substances, such as are used in percussion caps; and any preparation consisting of a mixture of a chlorate with phosphorus, or certain descriptions of phosphorus compounds, with or without the addition of carbonaceous matter and any preparation consisting of a mixture of a chlorate with sulphur, or with a sulphuret, with or without carbonaceous matter.

Division 2 comprises such substances as the chloride and the iodide of nitrogen, fulminating gold and silver, diazobenzol, and the nitrate of diazobenzol.

Class 6.—*Ammunition Class.*

The term "ammunition" means an explosive of any of the foregoing classes when enclosed in any case or contrivance, or otherwise adapted or prepared so as to form a cartridge or charge for small arms, cannon, or any other weapon, or for blasting, or to form any safety or other fuze for blasting, or for shells, or to form any tube for firing explosives, or to form a percussion cap, a detonator, a fog signal, a shell, a torpedo, a war rocket, or other contrivance other than a firework.

The term "percussion cap" does not include a detonator.

The term "detonator" means a capsule or case which is of such strength and construction, and contains an explosive of the fulminate-explosive class in such quantity that the explosion of one capsule or case will communicate the explosion to other like capsules or cases.

The term "safety fuze" means a fuze for blasting which burns and does not explode, and which does not contain its own means of ignition, and which is of such strength and construction and contains an explosive in such quantity that the burning of such fuze will not communicate laterally with other like fuzes.

The ammunition class has three divisions.

Division 1 comprises exclusively—

Safety cartridges,
Safety fuzes for blasting,
Railway fog signals,
Percussion caps.

Division 2 comprises any ammunition as before defined which does not contain its own means of ignition, and is not included in Division 1, such as—

Cartridges for small-arms, which are not safety cartridges,
Cartridges and charges for cannon, shells, mines, blasting, or other like purposes,
Shells and torpedoes containing any explosives,
Fuzes for blasting which are not safety fuzes,
Fuzes for shells,
Tubes for firing explosives,
War rockets,

which do not contain their own means of ignition.

Division 3 comprises any ammunition as before defined which contains its own means of ignition, and is not included in Division 1, such as—

Detonators,
Cartridges for small-arms, which are not safety cartridges,

Fuzes for blasting, which are not safety fuzes,
Fuzes for shells,
Tubes for firing explosives,

which do contain their own means of ignition.

By ammunition containing its own means of ignition is meant ammunition having an arrangement, whether attached to it or forming part of it, which is adapted to explode or fire the same by friction or percussion.

*CLASS 7.—*Firework Class.*

The term "firework" comprises firework composition and manufactured fireworks.

Division 1.—The term "firework composition" means any chemical compound or mechanically mixed preparation of an explosive or inflammable nature, which is used for the purpose of making manufactured fireworks, and is not included in the former classes of explosives, and also any star and any coloured fire composition, subject to the proviso herein-after set forth.

Division 2.—The term "manufactured firework" means any explosive of the foregoing classes, and any firework composition when such explosive or composition is enclosed in any case or contrivance, or is otherwise manufactured so as to form a squib, cracker, serpent, rocket (other than a war rocket), maroon, lance, wheel, Chinese fire, Roman candle, or other article specially adapted for the production of pyrotechnic effects or pyrotechnic signals or sound signals.

Provided that a substantially constructed and hermetically closed metal case, containing not more than 1 lb. of coloured fire composition, of such a nature as not to be liable to spontaneous ignition, shall be deemed to be a "manufactured firework."

---

* Order in Council No. 1A.

INTRODUCTION.

High and Low Explosives.—Explosives are, in popular language, divided into "High" and "Low," and of these two classes dynamite and gunpowder may be taken as the respective types. No hard and fast line can be drawn between them, but, speaking generally, we may class as "high" explosives those which are habitually fired by detonation, and used for purposes where a destructive rather than a propelling effect is aimed at. The effect of an explosion in a confined space is two-fold. It produces *pressure*, tending to simple rending of the containing envelope; and, rupture being effected, *work* or projectile effects.

Pressure and Work.—The *pressure* developed depends on the volume and temperature of the resulting gases or vapours, the *work* upon the heat evolved by the chemical changes involved in the decomposition of the original substance. This heat, which is not to be confounded with the simple heat of combustion, can be approximately calculated by thermo-chemical methods and data.

Maximum Work or Potential.—The maximum work which can be effected by a given chemical decomposition, of which an explosion is a particular instance, is given by the formula $E=425\ Q$, where E represents the maximum work or potential in kilogrammètres and Q the number of units of heat* evolved by that decomposition.

In practice only a fraction of this potential can be actually realised in the form of useful work, as a considerable amount is absorbed in heating the surrounding medium, in wave-making, and in various other ways. This fraction or modulus has been variously estimated in different cases at from 14 to 33 per cent. of the potential.

The formula given above shows that, *cæteris paribus*, the power of explosives may be compared by ascertaining the heat produced by their decomposition, provided that allowance be

---

* A unit of heat is the amount of heat required to raise one gramme of water from 0° to 1° C. In some cases heat is absorbed in a chemical change, and, of course, then E would have a minus value, but for obvious reasons this does not take place in the case of explosions.

made for dissipation of energy due to comparative slowness of action.

Theoretical Data.—In estimating theoretically the force of a given explosive we must know:—

(*a.*) Its chemical constitution and the equation of its decomposition.
(*b.*) The heat evolved in such decomposition.
(*c.*) The nature and volume of the gases and vapours formed by the decomposition.
(*d.*) The rapidity of the chemical reaction.

The chemical constitution, so far at least as regards the empirical formula, is a matter of simple analysis, and in cases where there is perfect combustion the nature of the resulting products may be predicted with considerable accuracy.

But when an explosive does not contain sufficient oxygen to ensure perfect combustion the equation of decomposition is not as a rule so simple, and the products of decomposition will vary considerably with the method in which it is effected. Hence, instead of representing the result of the reaction by one equation, several simultaneous equations may be involved, or the results may be so complicated as to resist expression by equations at all. It is, therefore, necessary to find out what is the final result of the reaction or series of reactions involved when a given explosive is fired in the way, and under the circumstances equivalent to the use to which it is desired to put it in practice. When we have obtained the equation or net result as above, the amount of heat evolved can be estimated from the data furnished by thermo-chemistry. The volume and consequent pressure of the gases and vapours can be calculated by well-known laws, assuming that such laws, known to hold good for ordinary temperatures, hold good also at the elevated temperatures obtained at the moment of explosion.* In the present state of knowledge we are not able to say whether this is the case or not, but it is probably true for the so-called permanent gases.

Thermal Values.—In estimating the thermal value of any chemical reaction it is only necessary to know the initial and

---

* This assumption is incorrect for two reasons. In the first place little is known about the specific heat of gases at high temperature, and, secondly, there is little doubt that dissociation takes place at high temperatures. The uncertainty with regard to these two points render it very difficult to estimate the temperature and pressure of explosion.—J. H. T.

final changes of the changing system. It is, indeed, a simple case illustrating the principle of conservation of energy.

For instance, the heat formation of a molecule of formic acid (CH_2O_2) is 99,420 units. The same amounts of carbon and hydrogen that are contained in that molecule when burnt in oxygen to form respectively carbonic anhydride and water give a total of about 165,320 units. Hence the heat evolved by burning the molecule in oxygen is 165,320 – 99,420 = 65,900 units.*

Dissociation.—From what has been said, it is apparent that if, as must often be the case, dissociation of the ultimate products of explosion takes place at the very high temperatures concerned, the total amount of heat evolved, or of potential, is not thereby effected. The tendency of dissociation is to diminish the initial pressure as less heat is evolved, but as temperature falls, this missing heat is reproduced in the union of the dissociated elements.

In short, if a given area represent the potential it is not altered by dissociation, but the configuration of its bounding line is altered much in the same way as the curve of pressure given by a slow gunpowder differs from that given by a quick one, though the total work may be the same.

Rapidity of Chemical Action.—The rapidity of chemical reaction is a very important factor. A piece of wood may slowly decay, and in the process evolve much the same ultimate products and amount of heat as when burnt in a furnace, but a boiler is not readily heated by letting wood slowly decay underneath it. The process is so slow that the heat or energy is dissipated as fast as it is evolved. Hence it is essential that the chemical changes involved in the decomposition of an explosive when practically applied should be extremely rapid.

To test the conclusions theoretically arrived at as to the power of an explosive, several methods have been suggested and tried. One very good practical test † is to bore vertical holes in cylindrical lead blocks, the latter being of as uniform material as possible, and of such dimensions as not to be disrupted by an explosion of a substance contained in the holes. Equal weights of different explosives are taken, placed in the bore-holes,

---

* *Manual of Chemistry.* Duprè and Hake. Vol. I., p. 72.

† Mr. Guttman informs me that the "lead test" was originated by Herr Trauzl, based on experiments by Captain Beckerheim in compressed tissue paper.

tamped with water or fine sand, and fired electrically or otherwise.

The relative effect of the different explosives is shown by the increased cavities in the blocks after firing. The cylindrical hole usually becomes a pyriform chamber, whose capacity is readily measured by determining the volume of water which it will contain and comparing this with its previously known capacity.

This method is not, however, readily applicable to slow explosives, as the tamping is blown away before the chamber is fully formed. Even in the case of high explosives variations are found depending on the physical condition of the explosive. For instance, rigidly frozen blasting gelatine gave considerably more effect than the same substance unfrozen. In the case of the latter, the bell-mouth form of the upper portion of the cylindrical hole showed that portions had exploded while actually passing out of the top of the hole.

Various modifications of crusher gauges have also been tried, notably the dynamometrical ring used by General Abbot, U.S.A., for ascertaining the relative effect in horizontal and other planes of various explosives immersed under varying depths of water. The results of his experiments are very striking, and forcibly point out the fact that the surrounding medium plays a very important part in the behaviour of explosives.

Indeed, speaking generally, it may be urged with much force that though theory may give very good guidance as to what may be expected of an explosive, and may satisfactorily explain apparent anomalies and eccentricities in its behaviour, yet nothing short of actual trial of that explosive under the circumstances and conditions under which it is proposed to use it, will give thoroughly dependable results as to its value.

To ask in a general way what is the best explosive for blasting is much the same as to ask what is the most useful tool in a carpenter's chest.

For instance, if in driving a tunnel heading it is found that with a given expenditure of time and money an explosive A gives a more rapid rate of progress than an explosive B, then clearly for that particular work A, *cœteris paribus*, is preferable. But it by no means follows that when the heading has got into another kind of rock A will maintain its superiority. When the object is simply to shatter the rock a quick and powerful explosive should be used; when, on the contrary, it is desired to obtain the rock in large blocks or masses, as in the case of slate or granite, then a slow explosive producing a rending, rather than a locally shattering effect, is preferable. Dynamite, for instance, is often extremely useful in sinking a shaft for coal, but when the coal itself is to be won, gunpowder, or, at all events, a slower explosive than dynamite, is commonly employed.

Practical Considerations applying to Explosives in general.—There are certain practical considerations which govern the choice and use of explosives in general, apart from any particular application of them.

The explosive should not be too bulky, for the less its density the greater will be the labour involved in boring holes to receive it. Moreover, a dense explosive will, other things being equal, evolve relatively more volumes of gas than a light and bulky one.

It should be reasonably safe to handle, transport, and store under ordinary conditions and when subjected to ordinary precautions, allowing for such amount of rough usage as it may fairly be expected to meet with. For instance, several explosives, powerful in their nature, and chemically stable, have been proposed for general use, which could be readily exploded by a wooden mallet or broomstick on a wooden floor. This was especially the case when the blow was of a glancing character, thereby combining friction and percussion. Such a blow is precisely what might be expected when charging a bore-hole, and for this reason such explosives have been considered too sensitive for general use. Many explosives containing chlorates become considerably more sensitive when left for a few months, especially if alternately exposed to moist and dry air.

Chemical stability, under lapse of time, and at varying natural temperatures, is essential, more particularly for Service explosives, which may be expected to be stored for a considerable length of time, and to be subjected to extreme atmospheric temperatures. The explosions of the blasting gelatine magazines at Aden in May and June 1888, furnish a very practical commentary on the importance of this point.

It is to be noted that a very small portion of impure or unstable material may, by its decomposition, ignite and finally explode large amounts of contiguous explosive, though this may be perfectly pure and stable in itself. The impure portion acts as a primer to the rest.

An explosive intended for use in mine galleries or other confined spaces should not give off actively poisonous gases or vapours, nor, so far as can be avoided, deleterious, either before or after explosion.

Practically, the explosion fumes of any explosive are more or less injurious in confined spaces, but some much less so than others.

Liquid explosives, or those which exude explosive liquids, are dangerous to convey or store owing to the risk of leakage, and are unfit for use in seamy ground, as some liquid may flow into a distant crevice unnoticed until itself accidentally struck by a boring tool, or otherwise exploded. Accidents are on record due to this cause in which the liquid had remained *perdue* for years.

In connexion with this point it may not be superfluous to add that all holes which have not "carried their burden" should be most carefully searched before again boring into or very near them.

An explosive which gives off acid or other corrosive fumes under ordinary conditions of storage is highly dangerous, as it may very possibly induce the spontaneous ignition of other explosives stored with it.

Distinction between Military and Civil Explosives.—At the same time a broad distinction is to be drawn between explosives used for emergent military purposes by experts, and those which are safe to trust in the hands of the ordinary miner. Occasions may well arise when an Engineer Officer would gladly avail himself of a powerful explosive which would be most objectionable for ordinary industrial purposes. Many such explosives could be readily manufactured in a very brief time from materials present in nearly every town as articles of commerce. In any calico printing establishment, for instance, a perfect mine of material for rough and ready, but powerful, explosives exists.

Effect of Mixtures of inert Substances with Explosives.—The effect of the admixture of an inert substance with an explosive varies considerably with their respective natures.

If the inert substance be of such a nature and be present in such quantity as to interpose between and insulate from each other the physical particles of the explosive, then the mixture will become much slower and less sensitive, or cease altogether to be explosible. As an instance of this may be cited the proposal of Mr. Gale to render gunpowder safe by mixing it with ground glass, which latter was to be sifted out when the powder was required for use.

Suppose a piece of honeycomb to represent a mixture of an explosive with inert matter. If the honey in the cells represents the explosive and the waxen walls the inert substance, then the mixture will be difficult to explode, at least by ordinary means. Such a state of things occurs when an inert liquid or melted solid is introduced into gun-cotton.

If, however, the circumstance be reversed, and the honey represents the inert matter and the wax the explosive, then we have a state of things like ordinary dynamite, in which each particle of kieselguhr absorbs, and is more or less surrounded by a continuous film of nitro-glycerine. As the continuity of the film is diminished, so the explosibility is lessened. Hence a mixture of kieselguhr with a very low proportion of nitro-glycerine is hardly an explosive; all the liquid is absorbed in the porous particles of the kieselguhr, and there is no connecting film. If, however, we substitute some non-porous substance, such as flakes of glass, or mica, for the absorbent, then each of the flakes will carry a surrounding film, and with a low proportion of nitro-glycerine we still have a powerful explosive.

The addition of a small per-centage of camphor or certain hydrocarbons to blasting gelatine or compressed gun-cotton considerably diminishes their sensibility, but the same addition to a discontinuous powder is said not to produce the same effect.

Sensibility of Explosives.—The sensibility of explosives to ignition or friction varies very considerably, and a low igniting point does not necessarily carry with it increased susceptibility to shock. For instance, fulminate of mercury fires at 190° C, oxalate of silver at 130° C, but the former is very much more easily exploded by friction than the latter.

The sensibility of a given explosive is considerably affected by its state of physical aggregation, as in the cases of compressed gunpowder and gun-cotton.

Temperature also plays an important part, for a very slight shock will induce the explosion of heated dynamite or gun-cotton, and even celluloid,* now much used for ornamental articles, will explode under a hammer when warmed to its softening point. Again, frozen dynamite or nitro-glycerine is comparatively insensible to a blow, while the reverse holds good with frozen blasting gelatine.

As a rule the sensibility of any chemical substance may be broadly stated to vary inversely with the temperature required

* Essentially a mixture of nitro-cotton and camphor.

to start decomposition, and with the specific heat, and directly as the heat evolved by the decomposition.

Safety in Coal Mines.—Recent accidents have caused attention to be directed to the question of safety of explosives for use in fiery coal mines. The sources of danger are twofold, viz:—firedamp, and fine coal dust, both of which are capable of forming an explosive mixture with air, and this mixture may be ignited by the act of blasting, under certain conditions which at present are not fully understood. Various proposals for rendering explosives flameless and for reducing the temperature of explosion have been made, and some of these will be found under the heading of Wetter Dynamite (p. 169). It is a matter of grave doubt whether any explosive exists which is incapable of igniting firedamp or coal dust under favourable conditions, but it has been pretty conclusively shown that nearly all of the so-called "High Explosives" are less dangerous than gunpowder in this respect.

Radius of Effect.—Professor A. G. Greenhill, F.R.S., was good enough to discuss with me, in October 1888, the question of the effects of explosion at a distance, and the results may be summed up as follows.

The sphere of gases suddenly formed on explosion will vary in radius as the cube root of the weight of explosive.

Setting aside the effects within this radius, which is practically "range of flash," Mr. Greenhill considers that the effect (as measured by the impulsive pressure) (E) of explosion varies as the square root of the weight (P) of explosive divided by the distance (D). Thus

$$E = m\frac{P^{\frac{1}{2}}}{D}$$

where m is a constant.

For example, if a given effect is produced at 1,000 yards by the explosion of nine tons—

$$E = m\frac{(9)^{\frac{1}{2}}}{1{,}000} = \frac{3\,m}{1{,}000}.$$

If the amount be 100 tons, and x be the distance at which the explosion of this amount will produce the same effect as nine tons at 1,000 yards, then

$$E = \frac{m\,(100)^{\frac{1}{2}}}{x} = \frac{m\,(9)^{\frac{1}{2}}}{1{,}000}.$$

$$\therefore x = 3{,}333 \text{ yards.}$$

But this applies strictly to cases where explosion takes place in an incompressible medium, *e.g.*, water at a great depth.

Mr. Greenhill thinks, however, that it may, for practical purposes, be applied to explosions in air.

It is hardly necessary to add that, in practice, the local accidents of the ground, the parts of ignition of the mass, and other circumstances, will greatly modify the above, and abnormal effects may often be produced by the evolution of vortex rings shot out from the envelope of the suddenly formed body of gases.

GUNPOWDER.

I do not propose to dwell at length on this explosive for the reason stated in the Preface. While during the past few years the manufacture and treatment of gunpowder for artillery purposes has received the closest attention, resulting in the production of powders differing most widely in physical characteristics from each other, and from the sweet simplicity which characterised the old powders in use some quarter of a century ago, the actual proportion of ingredients for black gunpowder has varied but little, if at all, and now, as then, the composition of nearly all descriptions of service gunpowder consists of the time-honoured 75 parts of saltpetre, 15 of charcoal, and 10 of sulphur.

While some Continental countries follow this formula, others somewhat vary the proportions.

Ordinary blasting powder* differs, broadly speaking, from the classes of powder in containing less saltpetre, and in being milled a shorter time, as a rule; also, less attention is paid to the charcoal used.

Briefly, the manufacture of gunpowder may be summarised as consisting of the following operations:—

1. Mixing the previously purified and sifted ingredients to form a "green," or "unworked" charge.
2. Milling (incorporating) the mixture to form mill-cake ("ripe," or "worked" charge).
3. Breaking down mill-cake. (This is omitted in many factories.)

---

* Blasting powder and gunpowder are sometimes spoken of by retail dealers as if quite different substances. They are, of course, identical, or, rather, blasting powder is simply a cheaper and inferior class of gunpowder. At the same time Desortiaux (p. 600) points out that not only is a powder rich in charcoal cheaper than ordinary powder, but that it may be expected to furnish a larger volume of gas when exploded, and he quotes Piobert in support of this theory. Hence, for blasting purposes, it is quite possible that a cheap powder, poor, comparatively speaking, in saltpetre, may be equal, or superior, to one of the ordinary composition.

4. Pressing.
5. Granulating, or "corning."
6. Dusting.
7. Glazing.
8. Drying in a stove.
9. Finishing, or final dusting.

In different factories the system employed in each of the above operations may differ in some degree, and in one or two cases the order of one or two processes may be changed, but the above is a sufficiently accurate general description.*

Of the ingredients mentioned, the charcoal is by far the most important, in one sense, for on its treatment depends much of the character of the gunpowder of which it forms a part. The material from which the charcoal is prepared, and the methods of carbonising this material, require most anxious attention, and the popular saying that "gunpowder all depends on the charcoal" is not far wrong.†

The properties of a given powder manufactured from given ingredients in given proportions depend on the physical characteristics of the finished powder, *e.g.*, length of milling, pressure given, amount of glazing, moisture, size and shape of grains or blocks, and their gravimetric density. These and other points go to make the art of making a really high-class powder suitable for a given task a matter of no small difficulty, as is shown by the care now found requisite in the manufacture, as compared with the somewhat haphazard methods employed some years ago.

It would be impossible, and, indeed, out of place here, to give at any length the history and manufacture of gunpowder, embracing the now numerous varieties of that explosive.

Nitrate Mixtures other than Gunpowder.

Of the explosives detailed in this class, a very large number are simply modifications of ordinary gunpowder, and differ from the latter only in the proportion of the ingredients, or by the addition of some other ingredients.

In many cases the main point is the substitution of other nitrates for the whole, or for a portion, of the nitrate of potash

* Of course, special varieties of powder, *e.g.*, prism powders, require exceptional details of treatment, *e.g.*, no glazing is required, as the prisms are pressed from grain powder.

† Pure saltpetre and pure sulphur obviously admit of no alteration by treatment consistent with retaining their purity, but charcoal may be made of innumerable varieties.

in ordinary gunpowder. Such nitrates are those of sodium, barium, and ammonium.

The following t ıble gives a comparison of the amount of oxygen contained in these bodies :—

—	Formula.	Per-centage of Oxygen.
		Per Cent.
Nitrate of soda (Chili saltpetre) - -	$NaNO_3$	56·47
„ ammonia - - -	NH_4NO_3	60·0*
„ potash (nitre, saltpetre) - -	KNO_3	47·48
„ baryta - - -	$Ba(NO_3)_2$	36·78

It would therefore, at first sight, appear to be highly advantageous to substitute one of the two first-named for saltpetre in gunpowder. Unfortunately, however, both of them are extremely hygroscopic, especially the ammonium nitrate, and consequently gunpowder made with them would, under ordinary conditions, soon become useless from damp absorbed from the atmosphere. In a hot, dry climate, nitrate of soda powders, especially if made only a short time before being required for use, would doubtless be valuable,† and be cheaper than gunpowder, and such powders were used to a considerable extent in the construction of the Suez Canal. Absolutely pure nitrate of soda is stated not to be unduly deliquescent, but the material as found in commerce contains other salts which are supposed to induce this property, and are difficult to remove by any reasonably economical process. The salt, however, is indirectly largely and increasingly used in the manufacture of gunpowder, for by the simple process of boiling it with chloride of potassium‡ it is converted into nitrate of potash, which is retained in the hot solution, while chloride of sodium is deposited. The principle of the process is the general rule that if two salts are dissolved together, which can form by

* But only one-third of this, viz., 20 per cent., is available as free oxygen, the remainder being required to form water, thus: $NH_4NO_3 = N_2 + 2H_2O + O$.

† It is stated on the authority of Berthelot that powder made with nitrate of soda is about one-third stronger than when made with saltpetre. (D. 605.)

‡ Chiefly obtained from carnallite ($KCl, MgCl_2, 6H_2O$), a mineral found in great quantity at Stassfurth in Saxony, lying over beds of rock salt. It resembles the latter in appearance, and contains 16 to 18 p.c. of potassium chloride. It is very deliquescent.

exchange of their metals a salt less soluble in the liquid, that salt will be deposited. Thus :—

$$NaNO_3 + KCl = KNO_3 + Na\ Cl.$$

The only apparent object of the use of nitrate of baryta is to produce a slow-burning powder, an end now generally attained by varying the physical characteristics of ordinary powder, notably as regards the size and shape of the grains.

Dr. Dupré has, however, pointed out that owing to the greater specific gravity of nitrate of baryta, this salt contains more oxygen for its bulk than nitrate of potash or nitrate of ammonia, and very nearly as much as nitrate of soda. This point is of importance in explosives for destructive purposes.

Nitrate of ammonia explosives have assumed considerable prominence of late, particularly in connexion with coal mining. The temperature of explosion of these mixtures is stated to be insufficient to ignite fire-damp or coal-dust. In most cases a nitro-compound such as nitro- or dinitro-naphthaline, benzol, &c., forms the other ingredient of the explosive, but in some instances a simple hydrocarbon or carbonaceous material is mixed with the nitrate of ammonia.

It should be mentioned that nitrate of ammonia alone is an explosive, though not a very powerful one.

The hygroscopic tendency of this salt is overcome by the use of a waterproof or hermetically sealed cartridge case. In some cases it has been proposed to waterproof the salt itself by impregnating it with melted hydrocarbons, &c.

Chlorate Mixtures.

The employment of chlorate of potash in explosive mixtures has always offered great attractions, owing to the violence of its effects due to the rapidity of its action. Hence the number of proposed explosives in which this substance forms the main ingredient is extremely large. At the same time they may, broadly speaking, be divided into two classes, those in which no particular attempt is made to diminish the dangerous sensibility of chlorate of potash compounds, and those in which by the addition of some diluting ingredient, or by some special mechanical treatment, endeavours are made to diminish this sensibility. Some of the compounds in the list appear to be simply more or less dangerous haphazard mixtures, while in other cases the same mixture of ingredients, or the same method of treatment, has with little, if any, variation, been proposed over and over again under different names. The objections to chlorate of potash compounds have been stated by several authorities.

For instance, Dr. Dupré, F.R.S., whose knowledge of explosives is exceptional in extent and minuteness, says:—"Chlorate of "potassium, on account of the readiness with which it lends "itself to the production of powerful explosives, offers a great "temptation to inventors of new explosives, and many attempts "have been made to put it to practical use, but so far with "very limited success. This is chiefly owing to two causes. "In the first place chlorate of potassium is a very unstable "compound,* and is liable to suffer decomposition under a "variety of circumstances, and under, comparatively speaking, "slight causes, chemical and mechanical. All chlorate mixtures "are liable to what is termed spontaneous ignition, or explosion "in the presence of a variety of materials, more particularly of "such as are acid or are liable to generate acid; and all "chlorate mixtures are readily exploded by percussion, but "more particularly by combined friction and percussion, such "as a glancing blow, which might easily and would often occur "in charging a hole. In the second place there is some evidence "to show that this sensitiveness to percussion and friction "increases by keeping, more especially if the explosive is "exposed to the action of moist and dry air alternately."†

The glancing blow referred to by Dr. Dupré is such as can be given with a slanting blow with a hammer on a plane or curved surface. Many chlorate mixtures which have been proposed for general use were found, on trial, to be so sensitive that they could be readily exploded by blows of this description with a light wooden mallet on a wooden floor, and almost all could be exploded with the same mallet on a stone floor.‡ The increased sensitiveness caused by keeping and exposure to moisture is probably due, in part at least, to the chlorate crystallizing out into fine crystals on the surface of the mixture.

Eissler, on the same subject, after referring to the fatal accident that attended the efforts of the discoverer of chlorate of potash, Berthollet, to utilise it in the manufacture of gunpowder,

---

* "Chlorate of potassium is one of those compounds during the decomposition of which heat is evolved, or energy is produced during decomposition, instead of, as is usually the case, being absorbed, and all such compounds are unstable."

† Annual Report of H.M. Inspectors of Explosives for 1885, p. 31.

‡ This practically reproduces the conditions of a wooden rammer in a stone bore-hole.

goes on to say:—"It is extremely doubtful from the peculiarities of this salt if anybody will ever overcome the "obstacles due to its inherent chemical properties, which nature "manifestly seems to have made unconquerable. In mixing "these compositions great danger is attendant, and too much "circumspection cannot be used. They explode instantly on "any violent stroke, very often by friction alone; sometimes "spontaneously, as when in a state of rest, and no known cause "for their combustion can be ascertained. Many are deluded as "to its safety by so-called experiments with freshly-made "powder. Manufacturers of the compound may attempt to "show its safety by hammering and cutting it, and similar "tests; but let the powder be exposed to the natural atmospheric action, attract some moisture during the damp foggy "night, then get dry, and the least friction or blow will cause "an unexpected explosion . . . . In ramming a cartridge "well home in a bore-hole the unsuspecting miner must either "use blows or pressure, the latter being equally dangerous from "the development of some friction. . . . ."*

Berthelot† remarks: "Les poudres au chlorate n'offrent donc "pas à ce point de vue (strength) sur les nouvelles matières "explosives une prepondérance qui puisse compenser les dangers "exceptionnels de leur fabrication et de leur manipulation. Ce "n'est que comme amorces que leur facile inflammation peut "offrir certains avantages."

Without going so far as to say that it is impossible to manufacture a safe chlorate mixture, it is a fact that out of many which have been examined with a view to their introduction into this country, not one has as yet been found to be safe enough to be licensed with the exception of *Asphaline*, but this explosive was not a practical success, inasmuch as its light and bulky nature required very large bore-holes in comparison with other explosives, and its manufacture and use in this country have been entirely abandoned.

Probably, also, the explosive papers such as *Melland's Paper Powder*, and *Reichen's Rolls* are fairly safe, but they have never, to my knowledge, come into practical use here, and the same objection of bulkiness in proportion to weight holds good with them as with asphaline.

I do not pretend to say that powerful and valuable explosives may not be, and have not been, manufactured with chlorate of

* Eissler, p. 139. † B., II., 325.

potash as their main ingredient, but I contend that though these are fairly safe when used for special purposes and by experts, none have as yet been brought to notice (with the previously named exception) which are suitable for general use by the mining population, and which could be relied upon not to cause accidents under ordinary conditions of transport, storage, and use.

In some cases it has been proposed to keep the ingredients of chlorate of potash mixtures apart till required for use, and this principle has been carried out in the case of *Rack-a-rock* and other explosives. *See* Sprengel Explosives, p. xli.

NITRO-GLYCERINE.

A nitro-compound may be defined as any chemical compound possessed of explosive properties, or capable of combining with metals to form an explosive compound, which is produced by the chemical action of nitric acid (whether mixed or not with sulphuric acid), or of a nitrate mixed with sulphuric acid, upon any carbonaceous substance, whether such compound is mechanically mixed with other substances or not.

It will be seen that this is a wide-reaching definition, embracing a vast number of substances, ranging from nitro-glycerine to picric acid and nitro-benzol. It includes many substances which, under the definition given in the Preface, are not ranked, at present at least, as explosives.

Nitro-glycerine was discovered in 1847* by A. Sobrero, in the laboratory of Pelouze, to be produced by the action of a mixture of strong nitric and sulphuric acids upon glycerine. It was known as pyro-glycerine, and subsequently as glonoine or blasting oil.

It was formerly considered as a nitro substitution compound, and the equation of its formation was given thus:—

$$\underset{\text{Glycerine.}}{C_3H_8O_3} + \underset{\text{Nitric acid.}}{3\,(H.\,NO_3)} = \underset{\text{Nitro-glycerine.}}{C_3H_5(NO_2)_3O_3} + \underset{\text{Water.}}{3H_2O},$$

which represents it as formed from glycerine by the substitution of three molecules of nitric peroxide (NO_2) for three atoms of hydrogen.

---

* Some authorities say 1846, Mr. Guttmann informs me: "It was expressly "stated to me by the late Mr. Sobrero that he made his invention in Turin, where "he was professor, and not in the laboratory of Pelouze. Sobrero gave the "name of pyro-glycerine, and in Avigliana there is still a small quantity kept, and "every year tested, of his original nitro-glycerine."

But more recent researches have shown that nitro-glycerine should rather be regarded as the nitric ester of glycerine. This view takes glycerine to be the triatomic alcohol of the compound radical glyceryl (C_3H_5), to which alcohol nitro-glycerine $C_3H_5(O.NO_2)_3$ has the same relation as nitric ethyl ester $C_2H_5(O.NO_2)$ has to ethylic alcohol (C_2H_5OH).

Under these circumstances the equation of formation should be represented thus :—

$$C_3H_5(OH)_3 + 3(HNO_3) = C_3H_5(O.NO_2)_3 + 3H_2O.$$

It will be noted that under both views the empirical formula of nitro-glycerine is the same, it is only the rational formula which is altered.

In the equations of formation no mention is made of the sulphuric acid, the presence of which is, however, essential in the production of the explosive. It takes, indeed, so far as we know, no active chemical part in the change, but is necessary to absorb the water which is formed in the process, and thus to keep the nitric acid up to its full strength and prevent its dilution, which would result in the formation of lower and weaker nitro-compounds of glycerine.

The following is a brief outline of the method of manufacture as carried out at the factory of Nobel's Explosives Company (Limited), at Ardeer, in the county of Ayr :—

A mixture of 1·2 tons of nitric acid (sp. gr. 1·5) with two tons of sulphuric acid (sp. gr. 1·84) is cooled down and run into a cooled leaden tank. Into this tank glycerine is injected in the form of fine spray till about 7·5 cwts. has entered the mixture. The temperature is very carefully watched, and never allowed to exceed a certain limit. The completion of this "nitrating" process is indicated by the fall of the thermometer due to the cessation of chemical action, and when the temperature has fallen to a certain point, the mixture of acids and nitro-glycerine is run off into another tank, where by virtue of the different specific gravities the nitro-glycerine (sp. gr. 1·6) separates in a short time, and floats on the top of the acids, whence it is drawn off, is well washed with water and with an alkaline solution to remove every trace of free acid, and is finally filtered into another tank ready for conversion into dynamite or similar explosive.

It is absolutely essential that the nitro-glycerine should be as pure as possible, free from all acid, and most especially from nitric peroxide (hyponitric acid, NO_2).

To attain this end, the ingredients must be pure, and the acids of the requisite strengths. The glycerine should be of 1·26 sp. gr. and free from lime, iron, and alumina, chlorides, and fatty acids. The presence of iron, alumina, or chlorine in any of the ingredients seriously interferes with the separation of the nitro-glycerine.*

By the chemical equation, for every 1 lb. of glycerine, 2·47 lbs. of nitro-glycerine should be furnished, but in practice the yield is only about 2 lbs., the loss being accounted for by the supposition that in addition to the production of the strongest form of nitro-glycerine, other nitro-compounds are formed, some of which are subsequently washed away.

Another method of obtaining nitro-glycerine is the Boutmy process, which was carried on in the Pembrey factory in South Wales in 1882, but abandoned in consequence of a serious accident. This process consisted in making two separate mixtures, viz., one, the sulpho-glyceric, of sulphuric acid and glycerine, and another, the sulpho-nitric, of sulphuric and nitric acids, and finally mixing the two together. The system had some advantages, but had the very serious disadvantage of slow separation, a considerable portion of nitro-glycerine being thus kept for a prolonged period in contact with the acids, thereby giving rise to danger. The yield by this process is only 185 to 190 of nitro-glycerine from 100 of glycerine.†

Other processes have been tried, but the method employed at Ardeer is, I believe, that most in use for practical purposes. Nitro-glycerine is a heavy liquid (sp. gr. about 1·6) of an oily appearance, varying from a colourless condition when quite pure, to a yellow or brownish yellow in some samples of the article when manufactured on a commercial scale. It is of a very sweet taste and without odour. It is an active poison, and mere contact with it will, in most persons, induce violent sickness and a specially painful form of headache. As a rule, however, custom soon reconciles most constitutions to it, and the workers in factories of it, after a short time, handle it all day without feeling any ill effects whatever. It is used in minute doses as a medicine, or rather as a palliative, in cases of *angina pectoris*, and finds a place in the British Pharmacopœia as a recognised drug. It explodes when heated to about 360° F.

* The above description is condensed from a portion of a lecture delivered by Mr. G. McRoberts, F.C.S., then works' manager at Ardeer, on 25th April 1883, and other points noted in this lecture will be also referred to in this chapter.

† D. 686.

or from a shock, but in small quantities it ignites and burns with some difficulty by the mere contact of flame.

When perfectly exploded, the resulting products are carbonic acid, nitrogen, water, and free oxygen, and may be represented thus:—

$$2\,C_3H_5(O.NO_2)_3 = 6\,CO_2 + 5\,H_2O + N_6 + O.$$

This excess of oxygen points to its profitable utilisation by mixing readily oxidisable substances with nitro-glycerine.

Imperfect combustion leads to the production of the poisonous carbonic oxide and nitrogen oxides, and hence the fumes of any nitro-glycerine compound, when simply burnt or imperfectly detonated, are far more dangerous than if the explosion be a perfect one.

The liquid freezes at about 40° F. as a rule,* though some samples freeze much less readily than others, possibly from their containing different nitro-compounds of glycerine. Once frozen it remains in this condition even when exposed for some time to a temperature sensibly exceeding its freezing point. When frozen it is very much less sensitive to a blow or to detonation than when liquid, and a detonator that will readily explode liquid nitro-glycerine, will simply scatter it when frozen. This holds good with most of the nitro-glycerine explosives, though there are some curious points of exception.

In a liquid state this explosive cannot be sold in, or imported into, this country. It is manufactured under the strict provision that it is forthwith made up into dynamite or some kindred licensed explosive.†

For many years after the discovery of nitro-glycerine, no practical use, except to a very limited extent as a medicine, was made of it, but between 1860 and 1863, Alfred Nobel, a Swedish engineer, whose name is indissolubly connected with, and pre-eminent in, the history of nitro-glycerine explosives, established factories on the Continent for its manufacture on a commercial scale as a blasting agent But in the then state of knowledge, the new explosive was not a complete success; it was poisonous

---

* The S.G. of frozen nitro-glycerine is 1·735, *i.e.*, it contracts $\frac{1}{12}$th of its volume.

† In America only is liquid nitro-glycerine, so far as I know, still used. Mr. Mowbray manufactured and used large quantities for the boring of the Hoosac Tunnel, and has published a very interesting volume on the subject. It is used chiefly for oil-well torpedoes at present in that country, but there, as well as in other countries, the cheaper and safer dynamites have replaced this sensitive and dangerous liquid form of explosive.

and dangerous to most people to handle, its liquid state rendered it dangerous to use and store, and it required strong confinement to even partially develop its power by simple inflammation. A great step was made when Nobel discovered the means of eliciting its enormous energy by the initial explosion of a detonator, which is simply a very powerful form of percussion cap. This marked a distinct epoch in the history of the explosive. In his English patent* he claims the ignition of the explosive by "the elevation of temperature produced by an " electric wire, or by an initiative explosion or detonation " produced in various ways, such as by a percussion cap or a " small quantity of gunpowder."

Without going minutely into the point of what is meant by detonation, it may be stated that it denotes the almost instantaneous resolution of a substance, or mixture of substances, into other substances, mainly or entirely permanent gases, which occupy a volume far exceeding that of the original substance or substances. In the case of nitro-compounds this detonation is effected by the combined action of a blow and heat, conveyed by the ignition of some very rapidly explosive and powerful substance placed as nearly as possible in contact with the substance to be detonated. Special explosives require special means of detonation, and it is curious that though an explosive *A* may detonate another explosive *B*, yet the action is not necessarily reciprocal, in that the explosive *B* will not necessarily detonate *A*. Sir F. Abel has shown that this holds good, at all events to a certain extent, with gun-cotton and nitro-glycerine.

The best and handiest method of detonation known for all nitro-compounds now in use, is to employ a stout metallic (usually copper) cap or detonator containing a charge of fulminate of mercury with, or without, the addition of chlorate of potash. The amount of the charge necessary to ensure detonation varies with the nature of the explosive to be detonated.

Returning to the history of nitro-glycerine, during the few years between 1863 and the finish of that decade, a succession of disastrous accidents due to liquid nitro-glycerine led to public reprobation of this dangerous and, to most people, mysterious and diabolical compound. Sweden, Belgium, and England (1869) totally prohibited its use, and Nobel's efforts and skill seemed destined to end in failure.

---

* Spec. No. 1813, 20.7.64.

To make nitro-glycerine safer in transport* and storage, Nobel proposed to dissolve it in twice its bulk of methylic alcohol or wood spirit (CH_3OH), thus producing an inexplosive liquid. The spirit and nitro-glycerine were easily separated when the latter was required for use, by simply pouring the mixture into water and stirring it. The nitro-glycerine sunk to the bottom, and could be obtained in separate form by decanting off the superincumbent layer of spirit and water.

There was some loss of nitro-glycerine in this process, which was at best but a cumbrous one, ill-adapted for general use in mining operations, and moreover the safety was apt to be illusionary, as the wood spirit is very volatile, and could evaporate freely from an imperfectly closed vessel.†

However, thanks to Nobel's energy and powers of invention, this method soon passed away as unnecessary, for in 1867 he at length produced dynamite, that is to say, a plastic mass consisting of nitro-glycerine absorbed in the pores of an inert material.

His English patent‡ is worth quoting at some length, as it not only marks the genesis of what has become an enormous manufacture, but more closely explains his views on the question of detonation. He says:—

"This invention relates to the use of nitro-glycerine in an altered condition, which renders it far more practical and safe for use. This altered condition of the nitro-glycerine is effected by causing it to be absorbed in porous unexplosive substances, such as charcoal, silica, paper, or similar materials, whereby it is converted into a powder, which I call dynamite, or Nobel's safety powder. By the absorption of the nitro-glycerine in some porous substance, it acquires the property of being in a

* Some at least of the catastrophes due to nitro-glycerine might have been averted had it not then been the practice to convey it in rigid metallic vessels. The danger of this was not then recognised, and another danger arose from the universal belief that when frozen, the liquid was much more sensitive than in its unfrozen condition. The belief was effectually dispelled from the mind of Mr. G. M. Mowbray, then the largest manufacturer of nitro-glycerine in America, by the failure of nitro-glycerine to detonate on an occasion when it had been accidentally upset in the snow in the winter of 1867–8. From that date he invariably sent out all his nitro-glycerine in a frozen state. But the belief died hard elsewhere.

† Another proposition, also by Nobel, was to dilute nitro-glycerine with half its weight of tar oil to render it inexplosive. When required for use 150 parts of the above mixture are shaken up with 100 parts of oleic acid, the latter dissolves the tar oil, setting free the nitro-glycerine. (Spec. No. 5,252, 28.4.85.)

‡ Spec. No. 1,345, 7.5.67.

high degree insensible to shocks, and it can also be burned over fire without exploding."*

He then speaks of the means of firing it, and goes on to say :—

"From the aforesaid it will be understood that a strong fulminating cap, if adapted to the fuze by being squeezed thereon, will cause dynamite to explode under all conditions of confinement or non-confinement. . . . It is evident that the above described fulminating cap may be greatly varied in form, but the principle for its action lies in a sudden development of a very intense pressure or shock."

Here then we have at last dynamite produced, and that name now stands as a generic title for a vast number of nitro-glycerine explosives, known individually under various distinguishing or fancy names. They may be conveniently divided into two great classes :—

1. Dynamites with an inert base acting merely as an absorbent for the liquid nitro-glycerine.
2. Dynamites with an active, that is to say an explosive or combustible, base.

This second class may be sub-divided into three minor classes, viz., those which contain as a base :—

(*a.*) Charcoal.

(*b.*) Gunpowder, or other nitrate or chlorate mixtures.

(*c.*) Gun-cotton, or other nitro-compounds.

Of Class 1, ordinary dynamite No. 1 may be taken as an example.

Of Class 2, dynamite No. 2, lithofracteur, and blasting gelatine may be taken as examples respectively of the sub-classes *a*, *b*, and *c*.

If it is desired to ascertain whether a given substance contains nitro-glycerine, the following are rough and ready tests :— If a liquid is oozing out or can be squeezed out from the substance, put the drop on to blotting paper. If this is nitro-glycerine it will make a greasy stain, not disappearing or drying away ; struck with a hammer on iron a loud report will be heard ; lit and burnt it will give a crackling sound and a greenish flame. If gradually heated by a flame underneath, it will give a sharp report.

Again, put a portion of the substance into a test tube and shake it up with methylic alcohol (wood spirit), first ascertain-

* As will be seen hereafter, this is true in a limited sense only.

ing that the spirit poured into water causes no turbidity or milky appearance. Filter the contents of the tube into another tube, and add pure water to the latter. If nitro-glycerine is present a milky appearance will be produced, and the heavy liquid will eventually settle at the bottom of the tube.

A much more delicate test is to use aniline and concentrated sulphuric acids as re-agents. In the presence of nitro-glycerine a purple colour is produced, which changes to green on the addition of water.

Nitro-glycerine explosives, unless carefully made from pure ingredients, are apt to decompose spontaneously. Any indication of acid fumes, or any tinge of green in them, should be followed by their prompt destruction, with suitable precautions.

NITRO-CELLULOSE.

To Braconnot, in 1832, is attributed the discovery of the conversion of starch, woody fibre, and similar substances, into highly combustible bodies by the action of concentrated nitric acid. The bodies so converted he called, generically, xyloidine. Pelouze six years afterwards pursued the subject, and extended his researches to cotton, paper, and vegetable substances generally. To him succeeded Dumas, who, with nitro-paper, which he termed nitramidine, proposed to make cartridges.

No practical result, however, came of these preliminary researches till Schönbein, in 1845, hit on the proper mode of manufacturing true gun-cotton by treating cotton with a mixture of strong nitric and sulphuric acids.

Nearly every country in Europe took up the matter with a view of utilising the new explosive for war purposes, and in Austria, under the auspices of von Lenk, the matter was closely investigated and the manufacture approved. The Austrians, indeed, in 1862, armed 30 batteries with gun-cotton cartridges in which the explosive was braided or twisted into yarns and ropes to diminish its rapid rate of burning. There, however, as in England and elsewhere, gun-cotton fell into disrepute from its unstable character, which resulted in many disasters.

Von Lenk had, it is true, introduced many improvements into the manufacture as distinguished from the crude methods hitherto pursued, but his project was not by any means perfect, and it was reserved for Abel to introduce his system of pulping, compressing, and thoroughly purifying gun-cotton, which has brought that explosive so much to the front, especially for military purposes.

The application of the principle of detonation to gun-cotton, and the discovery by the late E. O. Brown that perfectly wet and uninflammable (compressed) gun-cotton could be readily detonated by the detonation of a priming charge of the dry material in contact with it, raised it to the highest rank as a military explosive. As a mining agent it is to a considerable extent superseded by the plastic nitro-glycerine explosives, which are more easily inserted into an irregular bore-hole than the rigid gun-cotton.

In these pages I do not propose to enlarge on the manufacture and properties of gun-cotton, as they are fully given in the official text-books and elsewhere, but a few brief remarks may not be out of place.

Gun-cotton, or more generally nitro-cellulose, like nitro-glycerine, was formerly esteemed to be a nitro-substitution compound, but they are now both recognised as nitric esters, and the equation for the formation of gun-cotton (tri- or hexa-nitro-cellulose) is thus given, cellulose being taken as a hexatomic alcohol, $C_{12}H_{14}O_4(OH)_6$.

$$C_{12}H_{14}O_4(OH)_6 + 6\,HNO_3 = C_{12}H_{14}O_4(O.NO_2)_6 + 6\,H_2O.$$

This equation is precisely analogous to that given for the formation of nitro-glycerine.

The acid mixture employed in the manufacture of gun-cotton at Waltham Abbey consists of three parts by weight of sulphuric acid (S.G. 1·84) to one part of nitric acid (S.G. 1·52).

The process of manufacturing compressed gun-cotton at Waltham Abbey consists of nitrating* the previously cleaned and dried cotton, eliminating all the free acid, and pulping and compressing the gun-cotton into the required forms.†

To produce a stable product the greatest care is required in the selection of materials, and above all in the total elimination of free acid. It was want of attention to these points, or imperfect methods of obtaining these ends, that caused gun-cotton to fall into disrepute for so long.

---

* This word is commonly somewhat loosely used. To "nitrate" cellulose or glycerine means to convert it into nitro-cellulose or nitro-glycerine by the action of nitric acid, but "nitrated gun-cotton" means gun-cotton to which nitrates have been added. No real confusion need arise when the context is noted in each case.

† Full particulars of the manufacture will be found in the official text-book.

Theoretically 100 parts by weight of cellulose should produce 218·4 parts of gun-cotton, but practically, as in the analogous case of nitro-glycerine, the yield is a good deal less than this.*

The use of weaker acids results in the formation of collodion or soluble nitro-cotton, so called from its solubility in a mixture of ether and alcohol. This variety is said to consist mainly of penta-nitro-cellulose $C_{12}H_{14}O_4(O.NO_2)_5OH$, and tetra-nitro-cellulose $C_{12}H_{14}O_4(O.NO_2)_4(OH)_2$. The latter is often called di-nitro-cellulose, and its equation is given accordingly. No doubt lower nitro-compounds are present, but so far these have not been isolated.

Broadly speaking, gun-cotton and nitro-glycerine compounds have much in common as respects their behaviour when burnt, exploded, or detonated, or when exposed to shocks or blows. There is, however, an important difference between gun-cotton and such of the nitro-glycerine compounds as can be kept like gun-cotton in water. Wet gun-cotton absorbs water throughout its mass, and when thus saturated is not only absolutely uninflammable, but requires a very abnormal amount of fulminate to detonate it, though it is easily detonated, as previously stated, by the initial detonation of some of the dry material in contact with it. The case is different with (say) blasting gelatine, into which the water penetrates with difficulty. When taken out of the water it is readily inflammable or susceptible of detonation in the ordinary way. For mining purposes gun-cotton, or the nitrated forms of it usually employed in blasting, has the advantage of not freezing, nor can any liquid explosive exude from it. On the other hand its rigid character, as pointed out above, is not so suitable for loading the average more or less irregular bore-hole as a plastic material like dynamite and its congeners. But for many warlike purposes this very rigidity, which enables it when compressed or sawn into any required form to retain such form permanently, is very valuable.

The products of perfectly detonated gun-cotton may be fairly expressed by the following equation:—

$$C_{12}H_{14}O_4(O.NO_2)_6 = 9\,CO + 3\,CO_2 + 7\,H_2O + N_6$$

It appears then that gun-cotton does not contain sufficient oxygen to completely burn up its carbon, and hence in the

* As a matter of fact it is impossible on a large scale to manufacture gun-cotton which does not contain a considerable per-centage of lower nitrates.

varieties used in mining, some nitrate is always added to supply this defect, and this addition moreover tends to prevent the evolution of the actively poisonous CO.*

Here appears an additional reason why the blasting gelatines consisting of nitro-glycerine and soluble nitro-cotton form such a happy mixture. One has too little oxygen,† the other too much, and thus the two act as helpmeets to each other.

Picric Acid.

This class of explosives essentially consists of picric acid or picrates, and forms a convenient subdivision of the great nitro-compound class, other members of which have been previously enumerated.

Picric (carbazotic or tri-nitro-phenic) acid is a nitro-substitution compound obtained by the action of nitric acid on a variety of substances, *e.g.*, indigo, silk, acaroïd resin (*xanthorrhœa hastilis*), but on the commercial scale the substance now generally acted upon by the nitric acid is carbolic acid (phenol), and the equation of the process is simple, viz.:—

$$\underset{\text{Carbolic Acid.}}{C_6H_5.OH} + 3\ HNO_3 = \underset{\text{Picric Acid.}}{C_6H_2(NO_2)_3OH} + 3\ H_2O.$$

Picric acid may, as written above, be regarded as a picrate of hydrogen, which latter element can be displaced by a metal to form an ordinary picrate, *e.g.*, picrate of potassium $C_6H_2\ (NO_2)_3\ OK$. It is a crystalline substance of a brilliant yellow colour, and, as its name implies, is intensely bitter to the taste. It burns with a very smoky flame. It can be detected in very small amount in aqueous solution by heating with cyanide of potassium, when a red colour, due to the formation of isopurpurate of potassium, is produced.

When picric acid is burnt the fumes are very bitter, and give a peculiar bitter sensation at the back of the throat. This is a good rough test for even small quantities.

---

* I may here remark that it is not uncommon to hear that such and such an explosive gives off "no noxious fumes." No explosive exists to my knowledge of which this can be truly said. If *perfect* explosion or detonation be achieved, then we have at least a considerable amount of CO_2 which is not a harmless gas, but in actual work-a-day practice such perfection is exceptional, and we have CO and nitrous fumes evolved. Men have been killed by poisonous fumes from an explosive which was stated to be absolutely harmless in this way.

† The soluble nitro-cottons used in this explosive contain less oxygen in proportion than gun-cotton.

It is largely used as a dye, or constituent of dyes, and has not been usually considered as an explosive. Nor, indeed, does it usually behave like one under ordinary circumstances, though under special conditions easily produced it is capable of developing very formidable explosive properties.

It is somewhat sensitive to percussion, especially when warm.

A 1 lb. weight falling 26 inches on to picric acid on a steel anvil has exploded it. At 240° F. the height was 14 inches.

It may be burnt away in an unconfined state in considerable quantity without explosion,* but the mere contact of certain metallic salts or oxides with picric acid, in the presence of heat, develops powerful explosives, which are capable of acting as detonators to an indefinite amount of the acid, wet or dry, which is within reach of their detonative influence.

A disastrous explosion occurred in 1887† at a chemical manufactory near Manchester, originating with an accidental fire. This explosion was clearly due to the fulfilment of the conditions expressed in the last paragraph, and was the cause of certain experiments being made, which are specified at length in the report alluded to. The result of these was to show that picric acid could be readily detonated by the action of so small a quantity as five grains of fulminate of mercury applied in the usual way; and that such detonation would extend to wet picric acid, containing at all events 14 per cent. of water. In fact, exposed to detonation, the acid behaved much like compressed gun-cotton in sensibility and power of transmitting the initial detonation of some of the dry material to the same substance wetted.

No explosion or detonation could be obtained by simply heating or burning the acid; no very *large* amounts were tried, but

---

* Several authorities say that when sharply heated it explodes, but experience and experiment do not seem to bear this out.

† For full report on this, with details, see Colonel Majendie's special "Report to "the Rt. Hon. the S. of S. for Home Dept. on the circumstances attending a fire and "explosion at Messrs. Roberts, Dale, & Co.'s chemical works, Cornbrook, near "Manchester, on the 22nd June 1887," dated 15.8.87.

The dimensions of the Cornbrook craters were :—

East Crater.	West Crater.
30′ 4″ × 25′ 8″, and 8′ 3″ deep.	20′ 5″ × 22′ 8″, and 5′ 10″ deep.
(About 1,680 cubic feet.)	(About 710 cubic feet.)
Estimated amount (?) 7 cwt.	Estimated amount 130lbs.

The larger crater was estimated by Captain Friend, R.E., Secretary R.E. Committee, as about equivalent to that produced by 500 lbs. gun-cotton.

the burning away of so much as 1,500 lbs. harmlessly has been recorded. When strongly confined the continued application of heat exploded it.

A very crude admixture of metallic oxides or nitrates, notably litharge, lime, and the nitrates of lead, and strontium, with picric acid, will, on the application of heat, detonate, and such detonation will extend to contiguous masses of the unmixed acid. The heat first induces the formation of picrates, and its continued application causes their detonation.

In consequence of this accident, and of the subsequent experiments noted above, an Order in Council (No. 14, dated 29th December 1887) has been made under the 43rd section of the Explosives Act. The effect of this Order is to bring picric acid within the scope of the above Act, as regards its manufacture and storage, except when :—

(*a.*) It is wholly in solution.

(*b.*) When it is not wholly in solution, but is being manufactured or stored in a factory, building, or place exclusively appropriated to the manufacture or storage of picric acid, and in such manner as effectually to prevent any picric acid from coming into contact (whether under the action of fire or otherwise) with any basic metallic oxide or oxydising agent, or other substance capable of forming with picric acid an explosive mixture or explosive compound, or with any detonator or other article capable of exploding picric acid, or with any fire or light capable of igniting picric acid.*

Moreover, all picrates or mixtures of picric acid with any basic metallic oxide, &c., as detailed in (*b*), are to be deemed explosives within the meaning of the Act for *all*† the purposes of the said Act, unless such picrates or mixtures be wholly in solution.

---

* When, however, picric acid is manufactured with a view to being used as an explosive, it then falls under the scope of the Act for all purposes thereof in virtue of the provisions of sec. 3, quoted in the Preface, as being "used or manufactured "with a view to produce a practical effect by explosion." The exemptions in the Order apply to it when it is destined for other commercial purposes, *e.g.*, calico printing. A similar remark applies to picrates.

† *i.e.*, Not merely in case of manufacture and storage like picric acid.

In 1885 M. Turpin took out an English patent* claiming the employment of picric acid as an explosive agent for military and other uses unmixed with any ordinary substance whatever.† He proposed to compress the acid, to agglomerate and mould it with an aqueous solution of gum arabic, or oils, fats, &c.; or to use collodion jelly. When the last-named substance is diluted in the proportion of from 3 to 5 per cent. in a mixture of alcohol and ether, he states that the blocks of picric acid moulded with it will explode in a closed chamber with a priming of from one to three grammes of fulminate. He also proposed to cast picric acid into projectiles, the cast acid having a density of about 1·6. In shells he dispensed with the use of fulminate, and used, in lieu of it, a priming charge of violent gunpowder (25 grammes) or a powder composed of chlorate of potash, tar, and charcoal.

This invention probably closely resembles *Melinite,* but the exact composition of this explosive is kept secret.

Explosive mixtures containing picric acid with nitrates or chlorates are of very suspicious chemical stability, as the acid is a greedy one, and tends to form picrates displacing the chloric or nitric acids, especially in presence of moisture. Hence picrates are to be preferred to the acid itself for admixture with such other salts.

All the metallic picrates are explosive. The following is a list of those most likely to be proposed as ingredients in explosives:—

	Formula.
Picrate of ammonia	$C_6H_2(NO_2)_3NH_4O$.
" potash	$C_6H_2(NO_2)_3KO$.
" soda	$C_6H_2(NO_2)_3NaO$.
" baryta	$(C_6H_2(NO_2)_3)_2BaO_2 5H_2O$.
" lead	$(C_6H_2(NO_2)_3)_2PbO_2$.
" copper	$(C_6H_2(NO_2)_3)_2CuO_2 5H_2O$.

Picrate of ammonia is very feebly explosive, but has been used as an ingredient in Picric Powders. Picric acid also forms compounds with some of the hydro-carbons, but these are, of course, very deficient in oxygen.

---

* Spec. No. 15,089, 8.12.85.

† Picric acid is very deficient in oxygen, as its formula shows. The productions of its explosion will therefore largely consist of the actively poisonous carbonic oxide, and hence, as a blasting agent, in mines it would be objectionable. In digging out some shells which had been charged with some picric acid explosive and fired into earth, some French soldiers were poisoned by the noxious fumes some time after the shells had been fired and burst.

Other Nitro-compounds.

Numerous other nitro-compounds exist, the more prominent of which will be found under the prefix "Nitro," in the Dictionary. With the exception of nitro-cellulose and nitro-glycerine, the nitric ethers such as nitro-saccharose, nitro-starch, &c., are for the most part unstable. The true nitro-substitution products,* however, do not share this defect. The following is a list of those most usually proposed as ingredients for explosives.

—	Formula.	Nature of Substance.
Nitro-benzol	$C_6H_5(NO_2)$	Liquid.
Dinitro-benzol	$C_6H_4(NO_2)_2$	Solid.
Trinitro-benzol	$C_6H_3(NO_2)_3$	Crystalline solid.
Nitro-naphthaline	$C_{10}H_7(NO_2)$	Solid.
Dinitro-naphthaline	$C_{10}H_6(NO_2)_2$	"
Trinitro-naphthaline	$C_{10}H_5(NO_2)_3$	"
Tetranitro-naphthaline	$C_{10}H_4(NO_2)_4$	"
Nitro-toluol	$C_7H_7(NO_2)$	Crystalline solid.
Dinitro-toluol	$C_7H_6(NO_2)_2$	" "
Trinitro-toluol	$C_7H_5(NO_2)_3$	" "
Nitro-cumol	$C_9H_{11}(NO_2)$	Heavy liquid.
Dinitro-cumol	$C_9H_{10}(NO_2)_2$	Flaky solid.
Trinitro-cumol	$C_9H_9(NO_2)_3$	? Solid.
Nitro-phenol	$C_6H_4(NO_2)OH$	Crystalline solid.†
Dinitro-phenol	$C_6H_3(NO_2)_2OH$	Flaky solid.
Trinitro-phenol (picric acid)	$C_6H_2(NO_2)_3OH$	Crystalline solid.
Nitro-cresol	$C_7H_6(NO_2)OH$	Heavy liquid
Dinitro-cresol	$C_7H_5(NO_2)_2OH$	Crystalline solid.
Trinitro-cresol	$C_7H_4(NO_2)_3OH$	" "
Nitro-xylol	$C_8H_9(NO_2)$	Heavy liquid.
Dinitro-xylol	$C_8H_8(NO_2)_2$	Crystalline solid.
Trinitro-xylol	$C_8H_7(NO_2)_3$	" "

The prefix of "meta" to dinitro-benzol refers to chemical theories which need not be cited here, suffice it to say that

* The distinction depends on chemical questions, which need not be discussed here, though it may be stated that if gun-cotton, for instance, were really a nitro-substitution compound it might be expected to furnish an organic base under the influence of reducing agents, just as nitro-benzol under such influence furnishes the base aniline.

† Some of the mononitro-derivatives are sometimes produced in the solid and sometimes in the liquid form.

three series of nitro-benzols can be formed, the "ortho," "para," and "meta" series, which are identical in composition, but exhibit physical and other characteristics of difference between them.

As nitro- and dinitro-benzol are active poisons great care should be taken not to handle explosives containing them and unprotected by a cartridge case or other envelope, more than is absolutely necessary. On no account after such handling should food be touched with unwashed hands. Persons who are obliged to deal much with the bare explosive should drink milk.

Sprengel Explosives.

There are a considerable number of explosives which may all be classed under the patents of Dr. Herman Sprengel, F.R.S.* The essential principle of them all is the admixture of an oxidising with a combustible agent at the time of or just before being required for use, the constituents of the mixture being themselves non-explosive.

The idea of forming an explosive by the bringing into contact on the spot two inexplosive substances, or mixtures of explosives, is by no means novel, as may be seen on referring to the chlorate mixture section. It has indeed been proposed to make nitroglycerine on the spot where it was required, and the process proposed by Gale was simply reversing the order of things, when from his inexplosive mixture of gunpowder and ground glass he sifted out the explosive, leaving the inert glass behind. But Sprengel follows out a line of thought of his own, and concludes his introductory argument by saying—"Led by the "idea that (as a rule) an explosion is a sudden combustion, I "have submitted a variety of mixtures of oxidising and com-"bustible agents to the violent shock of a detonating cap. "These mixtures were made in such proportions that their "mutual oxidation and de-oxidation should be theoretically "complete." Some of the mixtures cited by him are liquid,

* No. 921, 6.4.71. No. 2,642, 5.10.71. *See* also Journal of Chemical Society, August and September 1873. Mr. Silas R. Divine, U.S.A., claims to have invented a mixture of chlorate of potash and nitro-benzol (now termed Rack-a-rock), and on 9th January 1871 he filed a caveat in the confidential archives of the "United States Patent Office," but published no patent till 1880. To Dr. Sprengel at all events would appear to belong the credit of the first publication, not of one explosive only of the type, but of the general principle underlying them all.

some are solid, others a mixture of a solid and a liquid. Amongst them are the following :—

(*a*) 1 chemical equivalent of nitro-benzol to 5 equivalents of nitric acid.
(*b*) 5 „ „ picric acid „ 13 „ „ „
(*c*) 87 „ „ nitro-naphthaline „ 413 „ „ „
(*d*) Porous cakes or lumps of chlorates of potash exploded—
(1) most violently with bi-sulphide of carbon.
(2) „ „ „ nitro-benzol.
(3) violently „ $\frac{1}{2}$ benzol + $\frac{1}{2}$ bi-sulphide of carbon.
(4) „ „ bi-sulphide of carbon saturated with naphthaline.
(5) very well „ carbolic acid dissolved in bi-sulphide of carbon.
(6) not well „ $\frac{3}{4}$ petroleum + $\frac{1}{4}$ bi-sulphide of carbon.
(7) „ „ benzol saturated with sulphur.
(8) not at all „ benzol alone.

On referring to the annexed alphabetical list it will be seen that—

(*a*) resembles *Hellhoffite.* Sprengel especially recognising the possible value of employing the dinitro-benzol.

(*b*) is *Oxonite.* Sprengel says, very truly, that picric acid alone is a powerful explosive when fired by a detonator.

(*d*) covers all the ground of *Rack-a-rock.*

When we come to the practical value of this class of explosives, whose great power cannot be disputed, we have to bring other considerations into play.

For instance, in this country no such explosives are, or can, in the existing state of the law, be licensed as "authorised explosives," supposing that they are to be made on the spot. For such admixture of ingredients to form an explosive for practical use, or for sale, would constitute "manufacture," and could be carried on only in a duly licensed factory. Even supposing this objection to be withdrawn, there would be serious objections to the use of such liquids as nitric acid or bi-sulphide of carbon in confined spaces, such as mine galleries, &c. In fact the use of such explosives would, as some critic has put it, "require a man to be a chemist as well as a miner."

All explosives of this class require a detonator. When nitric acid is one of the ingredients the greatest care is necessary to avoid its coming into contact with the explosive contained in the detonator. If such contact takes place a premature explosion is pretty certain, as indeed occurred in a trial with oxonite in August 1884.

Smokeless Powders.

The subject of smokeless powders has attained considerable prominence in the last few years, principally owing to the adoption of powders of this class for military use by all the European Governments.

Since the discovery of gun-cotton numberless attempts have been made to utilise this explosive for ballistic purposes. The first successful nitro-powders in this country were the Schultze and E.C. powders which were introduced for sporting purposes, and though not entirely smokeless, have attained a high degree of popularity.

The earliest nitro-powder for military purposes was produced in France towards the end of 1884, and since that time efforts have been made by every European Government to obtain a smokeless powder suited to their requirements. Though considerable secrecy has been maintained, there is little doubt that at the present time (1894) every leading European Power has adopted and is manufacturing on a large scale a powder of this description. Smokeless powders may be divided into two classes, A and B.

A.—Powders which consist of mixtures of nitro-glycerine with either soluble or insoluble nitro-cotton, other ingredients being sometimes added. The mixture is gelatinized by means of a solvent, usually acetone, or in the case of soluble nitro-cotton by the aid of heat. It is then squirted into cords, or rolled into sheets, and cut to grain of the required size. The finished material is generally of a tough leathery consistency. Examples of this class are Ballistite, Cordite, Maxim Powder, Amberite No. 2.

B.—Powders of which the principal ingredient is nitro-cellulose. This class may be subdivided into (*a*) powders which may be termed "semi-smokeless," *i.e.*, which contain nitrates of potash or baryta, and which must consequently always have a solid residue and give smoke (*e.g.*, Schultze, E.C. powders, &c.) and (*b*) powders which consist of gun-cotton only, or of which the products of combustion are wholly gaseous (*e.g.*, French and German Government powders, Walsrode, &c.). Various methods of manufacture have been proposed for powders of this class, but they all entail the gelatinization, either partial or complete, of the nitro-cellulose by means of a solvent. In the case of the powders of division (*a*), the material is usually only partly gelatinized, and then granulated by a suitable process. Those of division (*b*) are generally completely gelatinized and cut into small cubes or tablets.

FULMINATES.

Fulminates are compounds of the hypothetical fulminic acid ($H_2C_2N_2O_2$) and a metal. The most important explosive in this class is fulminate of mercury ($HgC_2N_2O_2$). It is produced by the action of nitrate of mercury on alcohol. It is incapable of being employed as an explosive by itself on account of its sensitiveness to percussion and friction. Its principal use is in the manufacture of percussion caps and detonators. The extreme suddenness of its action renders it most efficient in producing detonation in high explosives. Fulminate of mercury is always stored in water, and in this condition it is comparatively safe; but experiments have shown that even when moist it can be detonated by means of a small quantity of the dry material. Fulminate of silver ($Ag_2C_2N_2O_2$) is even more sensitive to percussion and friction than fulminate of mercury. It is used in minute quantities for crackers and other toy fireworks.

DICTIONARY OF EXPLOSIVES.

1. III $_{1}$ & $_{2}$. **Abel**, in 1865, patented his process of pulping and compressing gun-cotton (Spec. No. 1,102, 20.4.65), and in 1867 he patented gun-cotton in conjunction with a large portion of an oxydising body, such as chlorate or nitrate of potash, or nitrate of soda, or mixtures thereof, with the addition of a small proportion of alkali or of an alkaline carbonate such as carbonate of soda. He recommended a proportion of 70 to 40 per cent. of gun-cotton, with 30 to 60 per cent. of the oxidising substances, and added about 1 per cent. of the alkali. The same patent comprises the admixture of gun-cotton with nitro-glycerine. *See* GLYOXILINE. (Spec. No. 3,652, 24.12.67.)

2. III $_{2}$. **Abel** has also patented a smokeless powder consisting of nitro-cellulose pulp 100 parts and nitrate of ammonia 10 to 50 parts. The mixture is formed into blocks, cylinders, prisms or grains, which are rendered water-proof with petroleum or collodion. It is intended both for propulsive and blasting purposes. (Spec. 14,803, 14.9.86.)

3. **Abel.** *See* PICRIC POWDER.

4. **Abel.** *See* ELECTRIC FUZES and ELECTRIC TUBES.

5. III $_{2}$. **Acapina** resembles Schultze powder. It is made at Bologna for sporting purposes.

6. **Acetone** ($CH_3 . C_2H_3O$) is not an explosive itself, but is used as a solvent in the manufacture of smokeless powders and other explosives. It readily dissolves gun-cotton and nitro-glycerine.

It is obtained by the destructive distillation of calcium or strontium acetate.

Acetone is an ethereal liquid, lighter than water, and boils at 130° F. It burns with a luminous flame.

7. **Acetic Ether** ($C_2H_5 . C_2H_3O_2$) is obtained by distilling acetic acid with alcohol and sulphuric acid. It has been used as a solvent for gun-cotton in the manufacture of smokeless powder. It is volatile and has a pungent odour. It is apt to leave traces of acetic acid in explosives from which it has been driven off.

8. V_2. **Acetylides.** Some of the metallic acetylides possess explosive properties. Mercuric acetylide (3 C_2HgH_2O) is formed by the action of acetyline on mercuric oxide. Its properties have been studied by Travers and Plympton. It is in the form of a heavy white powder which detonates violently when suddenly heated or struck. (" Journal of Soc. of Chemical Industry," XIII. 277.)

9. IV_2. **Acme Powder,** invented by Mr. Liardet, consists of a mixture of picric acid, chlorate of potash, nitrate of potash, and tar. This explosive proved fatal to its inventor. It was submitted for license in the Colony of Victoria, and rejected. (Spec. No. 19,931, 23.10.93.)

10. III_2. **Actien-Gesellschaft Dynamit Nobel** have patented smokeless powders, consisting of nitro-cellulose, nitro-starch, or nitro-dextrin mixed with dinitro- or trinitro-toluol, benzol, xylol, or naphthaline. The mass is submitted to high pressure, and then granulated after the manner of black powder. (Spec. No. 6,129, 9.4.91.) (Fr. Spec. Nos. 212,649 and 212,650, 9.4.91.)

11. III_1. **Ætna Powder** is an American dynamite containing from 15 to 65 per cent. of nitro-glycerine, with wood pulp or nitrate of sodium. Some varieties contain roasted flour.

12. II. **Aix-la-Chapelle Powder** is so called from the place where it was tried. It consists simply of nitrate of soda and coal dust. (D. p. 609.)

13. V_1. **Alexander** proposed to employ in the manufacture of fulminating powder, 83 parts of amorphous phosphorus to 917 parts of nitrate of lead, or other suitable metallic salt. (Spec. No. 1,003, 9.4.57.)

Identically the same proposal was made by Johnson (Spec. No. 2,377, 10.10.56.)

14. III $_1$. **Allison Powder,** a porous gunpowder soaked in nitro-glycerine. (P. & S. 25.)

15. **A.** III $_1$. **Amberite No. 1** is a smokeless powder patented by Messrs. Curtis and André. The proportions of ingredients claimed as most suitable are as follows:—

Nitro-cellulose (insoluble) - -	40 to 47
„ (soluble) - - -	20 to 23
Nitro-glycerine - - -	40 to 30

To these may be added paraffin and shellac to modify the explosive force. In the earlier varieties, linseed oil was used in place of shellac for this purpose. The solvent used is either acetone or acetic ether, 50 to 60 parts being added to 100 of the explosive mixture. (Spec. No. 11,383, 4.7.91.)

16. **A.** III $_2$. **Amberite No. 2** consists of nitro-cellulose with or without nitrates, paraffin, vaseline, and graphite. Analysis of two samples gave:—

—	I.	II.
	Per Cent.	Per Cent.
Nitro-cellulose (insoluble) - - - - -	13·0	53·2
„ (soluble) - - -	59·5	24·1
Nitrate of barium and potassium - -	19·5	10·8
Paraffin - - - - - - - -	6·1	9·6
Volatile matters (chiefly water) - -	1·9	2·3

17. **American Powder.** *See* AUGENDRE.

18. III $_1$. **Americanite,** a term applied to an "insensitive nitro-glycerine" used by S. D. Smolianoff. It is a mixture of 80 per cent. of nitro-glycerine with a secret fluid mixture. (M. XV., p. 573, and XXI., p. 423.)

19. **A.** II. **Amide Powder** consists of charcoal, saltpetre or its equivalent, and ammonium salt in such proportions that on ignition a volatile and explosive amide compound is formed.

The proportions given are:—

Saltpetre - - - - - -	101 parts.
Nitrate of ammonia - - -	80 „
Charcoal - - - - - -	40 „

or in the form of a chemical equation:—

$$KNO_3 + H_4N\,(NO_3) + 3\,C = KH_2N + H_2O + CO + 2(CNO_2)$$

Amide Powder—*continued.*

The actual analysis of a sample gave :—

Saltpetre - - - -	49·64 per cent.
Nitrate of ammonia - - -	36·80 „
Charcoal - - - - -	12·67 „
Moisture - - - - -	0·89 „

The potassamide (KH_2N) is stated to be volatile and explosive at high temperatures, and to increase the useful effect of the powder. It is also claimed that very little, if any, residue is left, that no gases injurious to the gun are produced, and that there is much less smoke than with ordinary gunpowder. Amide powder has been found to ignite at 177° C when slowly heated on a sand bath, the igniting point for pebble powder similarly heated being 289° C and for Prism Brown 304° C.

20. **Amidogéne.** *See* AMMONIA DYNAMITE and GEMPERLÈ.

21. **R. III$_1$. Ammonia Dynamite** has several times been proposed for use as a powerful explosive, and no doubt it can lay claim to this designation. It essentially consists of nitro-glycerine mixed with nitrate of ammonia. Under the name of Ammoniakkrut it was introduced by Ohlson and Norrbin. Their form of this explosive consisted of finely powdered dry nitrate, or nitrite, of ammonia, intimately mixed with from 5 to 10 per cent. of its weight of powdered charcoal, coal dust, or other combustible substance. To the mixture was added 10 to 30 per cent. of nitro-glycerine. Picrate of potash, or nitro-mannite, were named as substitutes for the nitro-glycerine, but the latter was preferred on account of its liquid condition. (D. p. 721. T. pp. 91 and 102. Spec. No. 2,766, 18.9.72.)

In 1873 the British Dynamite Co. (now Nobel's Explosives Co.) submitted samples of ammonia dynamite to the Special War Office Committee on Gun-cotton, &c. These samples contained respectively :—

—	1	2
	Per Cent.	Per Cent.
Nitrate of ammonia - - - - -	75	70
Paraffin - - - - - -	4	7
Charcoal or coal dust - - - -	3	10
Nitro-glycerine - - - - - -	18	13

Ammonia Dynamite—*continued.*

They were accompanied by three other samples of much the same composition, save that in two of them nitrate of soda was used in place of nitrate of ammonia, and in the third, nitrate of potash took its place.

The Committee reported against the samples containing nitrates of ammonia and soda on the ground that, owing to the deliquescent nature of these salts, considerable exudation of nitro-glycerine might occur under the ordinary conditions of transport and storage. They reported in favour of the sample containing nitrate of potash (report of Spec. Com. on Gun-cotton, &c., 1871 to 1874, p. 96). The only feasible remedy against the danger here pointed out appears to be the enveloping of each individual cartridge in some absolutely waterproof covering.

22. III_1. **Ammonia Gelatine.** This explosive generally resembles the one last named, and is open, with some modifications, to the same objection. It consists of 40 parts of a thin blasting gelatine, *q.v.*, containing 97·5 parts of nitro-glycerine to 2·5 parts of nitro-cotton incorporated with 55 parts of nitrate of ammonia, and 5 parts of charcoal. It is a black, somewhat moist-looking mass, more friable than blasting gelatine, and less so than ordinary dynamite.

This explosive is open to the same objection which has hitherto barred the introduction of ammonia dynamite, that is to say, the liability of the deliquescent nitrate of ammonia to flow away in a liquid form, leaving only the thin blasting gelatine which would, from its physical characteristics, have a tendency to exude from an ordinary cartridge and form a source of danger. The explosive would in fact become a blasting gelatine "liable to exudation or liquefaction."

23. **Ammoniakkrut.** *See* AMMONIA DYNAMITE.

24. IV_2. **Ammonia-nitrate Powder** consists of:—

Ammonium nitrate - -	80 parts.
Potassium chlorate - - -	5 ,,
Nitro-glucose - - -	10 ,,
Coal tar - - - -	5 ,,

(M. No. XIII., p. 245.)

25. **Ammonic Powder.** *See* AMMONIA DYNAMITE.

26. III 2. **Ammonio-nitrate of Copper** is $4NH_3Cu(NO_3)_2$. It has been patented by Nobel as an explosive when detonated. It is said to be powerful and to give a very short flame of comparatively low temperature. (Spec. 16,920, 8.12.87.)

27. **A.** III 2. **Ammonite,** formerly called Miner's Safety Explosive, consists of nitrate of ammonia with mono- or dinitro-naphthaline. A sample gave :—

Nitro-naphthaline	- -	11·85 per cent.
Nitrate of ammonia	-	88·15 ,,

See FAVIER'S EXPLOSIVE.

28. **A.** VII 2. **Amorces** or **Toy Caps** are only toy fireworks, and consist of small dots of explosive composition enclosed between two pieces of thin paper (generally pink). The composition, as licensed in England, consists of a mixture of chlorate of potash and amorphous phosphorus, with or without the addition of nitrate of potash, sulphide of antimony, and powdered sulphur. The explosive is limited to 0·07 grains in each cap and the amorphous phosphorus component to 0·01 grains.

These, as above defined, are harmless toys, but if the amounts given are exceeded they are capable of exploding *en masse* and producing a disaster. At Vauves, near Paris, a child was cutting a toy cap with scissors, and thus caused the explosion of two packets containing 600 similar caps on the table. The child was killed. On the 14th May 1878, in the Rue de Beranger, Paris, six to eight million amorces, containing explosive to the amount of 154 grains per 1,000, exploded. The house and an adjoining one were destroyed; 14 people were killed and 16 wounded. The total amount of explosive was estimated at about 140 lbs.

A comparatively very small amount of amorces exploded on 3rd August 1888, in one of the sheds of a factory devoted to their manufacture near Wandsworth. Of four girls present, three were killed and the fourth severely injured. The light wooden shed in which the explosion occurred was utterly wrecked. There is no doubt that the accident was due to the detonation,

Amorces—*continued.*

by cutting with scissors or otherwise, of some amorces containing more than the stipulated amount of explosive. These acted as detonators to a number of loose amorces present. It was a parallel accident to the one at Vauves quoted above.

29. **Amsler.** *See* SCHENKER.

30. III$_1$. **Amylacé,** a mixture of anide powder with 40 per cent. to 68 per cent. of nitro-glycerine.

31. **Anders.** *See* HALOXYLINE and DIASPON.

32. III$_2$. **Anderson** dissolves nitro-cellulose in acetic ether diluted with 10 per cent. to 20 per cent. of benzol. The proportion is about two parts of liquid to one part of nitro-cellulose. The mass is agitated and then allowed to harden. (Spec. 13,308. 20.7.89.)

33. **André.** *See* CURTIS and ANDRÉ.

34. I. **Angular Powder.** A variety of blasting powder made in France. (P. & S. 48.)

35. V$_2$. **Aniline Fulminante** is a diazo-benzol compound, produced by treating nitrate of aniline ($C_6H_5\,NH_2 \cdot HNO_3$) with nitrous acid (HNO_2) prepared by the action of nitric acid on arsenic trioxide. The resulting products are water and diazo-benzol nitrate (aniline fulminante) having the composition $C_6H_5N_2 \cdot NO_3$. It crystalises out from ether and alcohol in long colourless needles, and is a very unstable compound, especially under the influence of moisture. When exposed to daylight it turns pink, and slowly decomposes. It is as sensitive to friction and percussion as fulminate of mercury. Heated to about 200° F. it explodes violently. It has been proposed for use in percussion caps, but is far too unstable.

36. **Antheunis.** *See* LITHOTRITE.

37. **Anthoine.** *See* PYROXILITE.

38. III$_2$. **Apyrite** is a gun-cotton smokeless powder used in the Swedish Navy (H. M.).

39. A. III_1. **Ardeer Powder,** a name given to dynamite containing sulphate of magnesia, with or without the addition of nitrates.

40. III_1. **Arlberg Dynamite** consists of 65 parts of nitro-glycerine absorbed in 35 parts of kieselguhr, barium nitrate, and charcoal.

41. III_1. **Asbestos Powder** is a combination of nitro-glycerine with asbestos as the absorbent. Other powders, such as gunpowder, white gunpowder, nitro-cellulose, &c., may be mixed with the compound, or with the asbestos alone. A portion of the latter may be replaced by clay, plaster, infusorial silica, chalk, &c. (T. p. 101.)

42. III_2. **Aspden** proposed a smokeless powder which, on examination, proved to be an imperfectly nitrated cotton, reduced to pulp. The sample submitted was strongly acid.

43. A.* IV_2. **Asphaline** consists of thoroughly cleansed wheat or barley bran, impregnated with chlorate of potash (in a proportion not exceeding 54 parts of chlorate to 42 parts of bran) mixed with saltpetre and sulphate of potash (in a proportion not exceeding four parts to 42 parts of bran). Paraffin oil, paraffin, ozokerit, and soap, or some of them, may be added. The compound is coloured pink with fuchsine.

A variety of the above is called Asphaline No. 2, and consists of the above, with the addition of saltpetre in such proportion that the total amount of saltpetre present in the explosive does not exceed 25 per cent. These explosives were licensed for manufacture in a factory near Llangollen, previously used for the manufacture of Pudrolithe, but the manufacture has been abandoned.

The explosive is easily recognised by its bulkiness and resemblance to red or pink bran. (Spec. No. 2,488, 8.6.81.)

44. III. **Atlas Dynamite,** patented by Kalk in 1883, consists of a mixture of nitro-glycerine, nitro-cellulose, nitro-starch, nitro-mannite, pyro-papier, and water-glass. (P. & S. 56.)

---

* No longer licensed.

45. III$_1$. **Atlas Powder** is a mixture of nitro-glycerine with wood pulp or (in some grades) sawdust. Nitrate of sodium is usually added. The per-centage of nitro-glycerine varies from 75 to 10 per cent. according to the brand. It is manufactured by the Repauno Chemical Company, Philadelphia, U.S.A., and is mainly known in England from its use in various attempted Fenian outrages. The following are examples of its composition :—

—	A.	B.
	Per Cent.	Per Cent.
Sodium nitrate - - - - - - -	2	34
Wood fibre - - - - -	21	14
Magnesium carbonate - - - - -	2	2
Nitro-glycerine - - - - - -	75	50

46. III$_2$. **Audemars** took the bark of mulberry or other trees of the *morus* genus, boiled it first with carbonate of soda and then with a solution of soap, washed it in hot water acidulated with nitric acid, and dried it by pressure. The fibre thus treated was soaked in a mixture of ammonia and alcohol and bleached with chloride of lime. It was then to be hackled, carded, and spun like cotton, treated with nitric acid and "con-" verted into an explosive compound resembling gun-" cotton." (Spec. No. 283, 6.2.55.)

47. **Audouin.** *See* EMILITE.

48. **Aufschlager.** *See* MÜLLER, WETTER.

49. IV$_2$. **Augendre's Powder,** known also as White German or American Powder, consists of :—

Chlorate of potash - -	50 parts.
Yellow prussiate of potash	25 „
Cane sugar - -	25 „

The materials are moistened, mixed in bronze mortars and granulated. (D. p. 614.)

50. **Azemar.** *See* SALA and AZEMAR.

51. II. **Azotine.** A mixture of nitrate of soda, sulphur, coal, and petroleum residue. It is manufactured in Hungary. (O. G.)

52. **A.** III 1. **Ballistite** is licensed as consisting of nitro-cotton combined with thoroughly purified nitro-glycerine, with or without the addition of camphor or aniline, in such proportions that the whole shall be of such character and consistency as not to be liable to exudation or liquefaction.

An original definition was: nitro-glycerine and nitro-cotton, with or without camphor, benzol, aniline, or similar substances, to be incorporated with nitrates, per-chlorates, or chlorates of potash, soda, or ammonia.

Practically, it is a blasting gelatine containing a large proportion of nitro-cotton, so as to render it slow enough to be used (as its name implies) as a propelling agent.

The latest definition also includes "such other sub-" stances as may from time to time be authorised by a " Secretary of State."

This explosive, as originally proposed by Nobel, was formed into sheets by passing the mixture between hot rollers, and was then cut into cubes. (Spec. No. 1,471, 31.1.88.) It has however latterly been sometimes incorporated with the aid of a solvent and squirted into cords. *See* CORDITE and FILITE.

53. III 2. **Bändisch Powder,** a variety of Schultze Powder.

54. IV 2. **Bantock** proposed to prepare nitro-cellulose by treating with the usual acids and adding a neutral salt. He suggests a mixture of 34 lbs. of nitric acid (S.G. 1·5), 65 lbs. of sulphuric acid (S.G. 1·84), and 1 lb. anhydrous sulphate of potassium. To 100 lbs. of this mixture 8 lbs. of dry cellulose is to be added. To the nitro-cellulose thus prepared 25 lbs. of saltpetre and 15 lbs. of chlorate of potash are to be added. As will be seen this (with the exception of the addition of the neutral salt) is simply Abel's nitrated gun-cotton. (Spec. No. 4,806, 12.12.76.)

55. III 2. **Barbe** proposed to render gun-cotton less sensitive and inflammable by adding nitrates, principally nitrate of ammonia. (Fr. Spec. No. 159,214, 17.12.83.) He also patented the addition of carbonate of ammonia and such like bodies to explosives to increase their stability. (Fr. Spec. No. 168,189, 10.4.85.)

56. III_2. **Barnwell** proposed to use pyroxiline in solution as a varnish, or in combination with plastic substances for moulding. In a powdered form it was proposed to use it in gunpowder as a substitute for charcoal. (Spec No. 2,249, 15.9.60.)

57. IV_2. **Baron** and **Cauvet** submitted to the French Government Commission, in 1882, two powders, consisting of:—

—	No. 1.	No. 2.
	Per Cent.	Per Cent.
Chlorate of potash - - - - - -	50	50
Prussiate of potash - - - -	50	25
Sugar - - - - - - -	—	25

They are simply varieties of Augendre's powder.

58. II. **Barytic Powder** consists of a mixture of eight parts of gunpowder with two parts of nitrate of baryta powder. It was used in Prussia (1865) for heavy guns.

59. **Bautzen* Powder** consists of equal parts nitrated wood and nitre. (O. G.)

60. IV_2. **Bayon** has patented a mixture of chlorate of potash, gum Arabic, and coarse bran. The mixture can be used in grain or in the form of cartridges. (Fr. Spec. No. 144,903, 20.9.81.)

61. **Beadle.** *See* Cross.

62. **Bela de Broncs.** *See* Bronslithe.

63. IV_2. **Bellford's Powder** consists of:—

Charcoal - - - - - -	19·5 parts.
Saltpetre - - - -	68·8 ,,
Sulphur - - - - - -	12·5 ,,

mixed, pressed, and granulated. The powder so produced is impregnated with a saturated aqueous solution of chlorate of potash, dried at 100° F. for four days, and used unglazed. (Spec. No. 2,910, 15.12.53.)

64. **A.** III_2. **Bellite,** invented by C. Lamm of Stockholm, resembles Roburite, Securite, and numerous other explosives. Its composition is as follows :—

—	No. 1.	No. 2.
	Per Cent.	Per Cent.
Meta-dinitro-benzol - - - - - -	15	34
Nitrate of ammonia - - - - -	85	66

65. III_1. **Bender** has patented an explosive called "Compressed Grisoutine." It is made by compressing dynamite containing nitrate of ammonia or other active salt with or without an inert absorbent. (Fr. Spec. No. 208,200, 12.9.90.)

66. V_1. **Benedict's Powders** are proposed for use in caps instead of fulminate of mercury. They are "single" and "double," and have respectively the following composition :—

—	Single.	Double.
	Parts.	Parts.
Chlorate of potash - - - - - -	12	9
Amorphous phosphorous - - -	6	1
Lead oxide - - - - - - -	12	—
Rosin - - - - - - - - -	1	—
Sulphide of antimony - - -	—	1
Sublimed sulphur - - - - - -	—	0·25
Saltpetre - - - - - -	—	0·25

(M. No. V., p. 752.)

67. IV_2. **Bengaline,** proposed by Medail in 1882, consists of bran (three parts), steeped in a solution of chlorate of potash (two parts). It is employed in the form of compressed cartridges. *See* ASPHALINE.

68. II. **Bennett's Powder** consists of the ordinary ingredients of gunpowder, to which a proportion of slaked lime, gypsum, or good cement is added. The mixture is made into a hard paste, which is subsequently granulated.

The proportions given are :—

Saltpetre - - - - - -	65 parts.
Sulphur - - - - - -	10 ,,
Charcoal - - - - -	18 ,,
Lime - - - - - -	7 ,,

(D. p. 600. Spec. No. 3,206, 21.12.61.)

69. **Benzo-glyceronitre.** *See* HEUSSCHEN (432).

70. IV_2. **Berg** and **Carimantrand** propose a mixture of chlorate of potash and hypophosphite of barium or sodium.

71. III_2. **Berg Roburite** consists of dinitro-benzol and nitrate of ammonia, with or without the addition of phenol (Salvati).

72. **Berg.** *See* NITROLKRUT.

73. **Bergenström.** *See* SALITE.

74. IV_2. **Berthelot,** an explosive submitted for license in the Colony of Victoria and rejected. It consists of :—

Chlorate of potash - - - -	80 per cent.
Vaseline - - - - Paraffin - - - -	} 10 „
French chalk - - -	10 „

75. **Bevan.** *See* CROSS.

76. III_1. **Bichel** distils hydro-carbons, such as resin oil, wood, or coal tars, with sulphur, and mixes the resulting product with nitrates or chlorates, or nitro-compounds. He claims that the admixture with nitro-glycerine, &c., is readily made, and results in a stable compound. (Spec. No. 14,623, 11.11.86, amended 5.10.88.)

77. III_1. **Bichel** has patented the use of explosive liquids containing nitric acid, mixed with fossil flour, in cartridges formed of sheet-lead or lead-tin alloy. (Fr. Spec. 171,169, 14.9.85.)

78. **Bickford.** *See* NITROFLAX.

79. VI_1. **Bickford Fuze.** *See* SAFETY FUZE.

80. **A.** VI_2. **Bickford's Patent Volley Firer** is an arrangement for firing a number of shots simultaneously. It consists of a cylinder of metal or other material, containing a priming paste of mealed powder, into which the ends of instantaneous fuzes are fitted. One length of safety fuze ignites the arrangement.

81. IV_2. **Bjorkmann, C. G.,** proposed an explosive consisting of the following:—

Nitrate of potash - - -	20 per cent.
Chlorate ,, - - - -	20 ,,
"Cellulosa" - - - -	10 ,,
Peameal - - - - -	10 ,,
Sawdust - - - -	10 ,,
"Nitroline" - - - -	30 ,,

The "cellulosa" is made by the action of 20 parts nitric and 40 parts sulphuric acid on 12 parts peameal.

The "nitroline" is made by the action of 80 parts nitric and 170 parts of sulphuric acid on 15 parts raw stearic oil and 15 parts syrup.

The compound is fired by a priming charge of gunpowder. (T., p. 104)

It is obvious that the two explosives Nos. 81 and 84 are modifications of the same. The first description is from the American, the second from the English patent.

82. III_1. **Bjorkmann, C. G.,** proposed also to mix glycerine with one-third of its weight of a carbo-hydrate, such as sucrose or glucose, and to heat the mixture with 2½ times its weight of strong nitric acid. This "blasting oil," he claims, is not nitro-glycerine.*

He proposes an explosive composed of:—

"Blasting oil," as above - -	60 per cent.
Brinoxide of manganese - -	18 ,,
Prussiate of potash - - -	10 ,,
Sulphide of antimony - - -	2 ,,
Pine sawdust or coal dust - -	10 ,,

The above is stated not to freeze, to be insensible to shock, and not to be poisonous. (Spec. No. 2,483, 19.6.80.)

83. **Bjorkmann, C. G.,** *See* KRAFT.

84. IV_2. **Bjorkmann, E. A.,** treated sugar or other saccharine matters with nitric and sulphuric acids and added 25 to 50 per cent. of this, which he termed "Nitroline," to nitrate and chlorate of potash, cellulose, charcoal, other vegetable substances, coal, tannin, or "compounds of such substances." The explosive could be made as a powder or of the consistency of wax, and was called Vigorite. (Spec. 2,459, 8.7.75.)

* But *see* GLUKODINE.

85. I. **Black Powder** is ordinary gunpowder.

86. III$_2$. **Blasting Amberite** consists of Amberite No. 2, with the addition of woodmeal.

87. **A.** III$_1$. **Blasting Gelatine.** This essentially consists of a combination of nitro-glycerine and nitro-cotton. It is manufactured by dissolving finely-divided soluble nitro-cotton in nitro-glycerine, heated to a point (about 100° F.) below that at which it begins to decompose. As nitro-glycerine contains an excess, and nitro-cotton a deficiency, of the oxygen requisite for the complete combustion of the carbon contained in them, blasting gelatine, by combining the two, compensates for the respective surplus and defect of available oxygen.

Two varieties are licensed in this country, viz., No. 1, which is defined as " nitro-cotton . . . . combined " with thoroughly purified nitro glycerine in such pro- " portions that the whole shall be of such character and " consistency as not to be liable to liquefaction or " exudation."

No. 2 is simply No. 1, with the addition of a nitrate, with or without charcoal.

The No. 1 is that practically in use. It is a gelatinous mass, looking something like new honey in colour, varying in consistency from a tough leathery material to a soft one, like ordinary stiff jelly. These variations depend on a variety of circumstances, such as the chemical condition of the nitro-cotton, and certain other details of manufacture. It contains from 93 to 95 per cent. of nitro-glycerine, and is issued in cartridges like dynamite. Speakly broadly, the thinner the gelatine the more sensitive it is to detonation, but on the other hand, a thin gelatine is more liable to liquefaction, and possibly also to exudation, and thus to cause danger in storage and transport. Specially strong detonators are required to explode blasting gelatine, or ordinary detonators with a primer of dynamite or gunpowder. It requires confinement to develop its detonative power, or rather its power of transmission of detonation, for a train of it cannot be exploded in the open (unlike dynamite) except by means

Blasting Gelatine—*continued.*

of a very powerful initial detonation. The addition of a small per-centage of camphor or other substances soluble in glycerine and rich in carbon and hydrogen, as benzol or nitro-benzol, renders it very insensible to explosion by shock or blow, and hence more difficult to detonate. While dynamite and nitro-glycerine in a frozen state are much less liable to be exploded by a blow, such as that given by a rifle bullet, than when unfrozen, the reverse holds good with frozen blasting gelatine.

Some recent accidents tend to show that considerable care should be exercised in charging bore-holes with wholly or partly frozen gelatinous compounds of nitro-glycerine. No undue force or friction should be permitted, even with a wooden rammer.

This explosive has the great advantage of being, unlike dynamite, practically entirely unaffected by water, and so can, if desirable, be kept under water like wet gun-cotton. It is on the whole less liable to freeze than dynamite.

These properties, combined with its great strength, due to the increased proportion of nitro-glycerine and the use of the explosive nitro-cotton instead of the inert and non-carbonaceous kieselguhr, render it a formidable rival to ordinary dynamite. It is moreover more cleanly to use, and, from its physical condition, small portions are not so liable to be broken off the main cartridge by accident and to fall into places where their presence may constitute a serious hidden danger.

It is most essential that the greatest care be taken in the manufacture of blasting gelatine, otherwise it is apt to seriously deteriorate and even to decompose altogether in store, but when well made and from pure ingredients it appears to be a safe and stable explosive.

It is issued in cartridges and packed like Dynamite No. 1, *q.v.*

For a paper on the manufacture of this explosive, by Mr. McRoberts, F.C.S., see *Jour. Soc. Chem. Ind.*, Vol. IX., p. 265 (March 1890).

88. A. **Blasting Matagnite.** *See* MATAGNITE.

89. II. **Bleckmann's Powder** (also called Haloxyline) consists of sawdust freed from resinous matter, or other powdered cellulose substances, saltpetre, and charcoal, with the occasional addition of ferrocyanide (yellow prussiate) of potash, when a quick explosion is required.

The proportions are:—

Sawdust	9	parts.
Charcoal	3 to 5	,,
Saltpetre	45	,,

The substances are mixed, moistened with one pint of water to about 112 lbs. of the mixture, crushed and ground in a mill, made into cake, and treated like ordinary gunpowder. (Spec. No. 1,341, 10.5.66.)

90. A. III_2. **B. N. Powder** is the name given to a semi-smokeless powder manufactured in France. It consists of partially gelatinised nitro-cotton incorporated with tannin and nitrates of baryta and potash, and contains about 2 per cent. of normal moisture. It is made in flat strips or in tablets of a drab colour, hard and brittle. (P. & S. 105.)

91. III_2. **Bobœuf Powder** is similar to Designolle's Powder, *q.v.*

92. **Boghead.** *See* DYNAMITE DE BOGHEAD.

93. **Böhm.** *See* FALKENSTEIN.

94. IV_1. **Bolton, Sir F.**, proposed a mixture of chlorate of potash, or other chlorate or nitrate with nitro-benzol or other solvent in which is dissolved a carbonaceous material such as resin, molasses, &c. He proposed to make cartridges by filling bags with the solid constituent and then saturating them with the liquid, just as in the case of Rack-a-Rock, *q.v.*

95. II. **Bolton's Powder** consists of:—

Carbonate of copper	8	parts.
Graphite	10	,,
Prepared quicklime	14	,,
,, alum	50	,,
,, sugar	350	,,
Nitrate of soda	350	,,
Soda ash	20	,,

Bolton's Powder—*continued.*

Ferrocyanide of potassium	-	300 parts.
Charcoal	-	30 „
Carbonate of potash	-	450 „

The carbonate of copper, lime, nitrate of soda, soda-ash, and carbonate of potash are mixed with one-half of the graphite and charcoal to form one mixture, the remainder of the ingredients with the remainder of the graphite and charcoal to form another mixture. Each of these is stated to be inexplosive and harmless. When required for use, the two mixtures are to be combined. (Spec. No. 342, 31.1.68.)

96. **Boritine.** *See* TURPIN (981).

97. **Borland.** *See* CARBO-DYNAMITE and JOHNSON BORLAND POWDERS.

98. III$_2$. **Borlinetto's Powder** consists of:—

Picric acid	-	10 parts.
Nitrate of soda	-	10 „
Chromate of potash	-	8½ „

It is said to be insensible to friction and percussion. (D., p. 736.)

99. **Bornhardt.** *See* ELECTRIC DETONATOR FUZES.

100. III$_1$. **Bouchard-Praceiq** has patented a method of manufacture of nitric acid, and the improvement of gunpowder by substituting for the whole or part of the sulphur, hydrated nitric acid. (Fr. Spec. 197,358, 12.4.89.)

101. V$_1$. **Bousfield** proposed fulminate of mercury and collodion as a detonating compound. (Spec. No. 2,882, 17.11.57.)

102. I. **Bowen** proposed to make gunpowder from a mixture of 18 to 21 parts of carbonised and ground lignite, four parts of sulphur and 74 to 77 parts of saltpetre. (Spec. No. 3,876, 9.8.83.)

He also used charcoal made by carbonising maize and other cereals. (Spec. No. 3,953, 13.3.86.)

102^a. II. **Boyd** proposes the following mixture:—

Oxide of iron	-	4 parts.
Nitrate of potash	-	4 „
Nitrate of baryta	-	1 „
Kerosene shale	-	3 „
Sulphur	-	2 „
Wood-meal	-	2 „

(Spec. No. 24,425, 19.12.93.)

103. **B. P. Powder.** *See* PLASTORNENITE.

104. III$_2$. **Bracket's Sporting Powder** has the following compositions by analysis:—

Soluble nitro-lignin - - -	31·43 per cent.
Insoluble ,, - -	13·70 ,,
Lignin (charred) - - -	13·22 ,,
Humus - - -	18·94 ,,
Sodium nitrate - - -	19·76 ,,
Moisture - - -	2·95. ,,

(Munroe.)

104[a]. **Bradbury.** *See* HARRISON.

105. **Brady.** *See* VULCAN.

106. IV$_2$. **Brain's Powders** consist of 40 to 50 per cent. of nitro-glycerine mixed with 60 to 50 per cent. of other substance, as chlorate of potash, sugar, charcoal, ground coal, sawdust, dextrine, starch, and shumach. The last-named substance consists of the ground shoots of the *Rhus Coriaria*, and contains a large proportion of tannic acid. It is used in tanning morocco leather. The following are examples of these powders, 40 to 50 per cent. of nitro-glycerine being added in each case:—

——	1.	2.	3.	4.	5.
	Parts.	Parts.	Parts.	Parts.	Parts.
Chlorate of potash - - -	15	1	1	1	1
Charcoal or sawdust - - -	10	1	2	1	1*
Ground coal - - -	5	1	—	—	1†
Sugar - - - - -	—	—	—	1	—
Saltpetre - - -	—	—	—	—	1

(Spec. No. 2,984, 11.9.73.)

107. **Brain.** *See* ELECTRIC FUZES.

108. II. **Brandeisl's Powder** consists of 16 parts saltpetre, 2 of sulphur, and 3 of sugar.

109. **Brank.** *See* VON BRANK.

110. **Brise-rocs.** *See* ROBANDIS.

* Sawdust, dextrin, starch, or shumach. † Ground coal or charcoal.

111. II. IV_2. **Britainite,*** a name given to an explosive invented by Dahmen. It consists of nitrates, chlorates, and resin. The nitrates and chlorates are thoroughly dried, and then incorporated in a ball mill, with a solution of resin, colophony, hard pitch, or asphalt. The solvent is evaporated, and the residue is then ground. (Spec. 15,566, 16.8.93.)

A sample examined, however, gave the following composition :—

Nitrate of ammonia - -	70·80	per cent.
Nitrate and chlorate of potash -	20·40	,,
Naphthaline - - -	7·10	,,
Moisture - - - -	1·70	,,

111a. **Brodersen.** *See* GLYOXILINE.

112. **Broncs.** *See* BRONOLITHE.

113. **A.** III_2. **Bronolithe,** the invention of M. Bela de Broncs, consists of various mixtures, of which the principal ingredients are picrates of lead and sodium and potassium, with the addition of nitro-naphthaline and soot.

This seems to be similar to a series of explosives noted by Munroe (No. XIII., p. 247) as consisting of :—

Barium-sodium picrate - -	15 to 30	per cent.
Lead-sodium- - - -	8 ,, 30	,,
Potassium - - -	2 ,, 10	,,
Nitro-naphthaline - - -	5 ,, 20	,,
Saltpetre - - - -	20 ,, 40	,,
Sugar - - - -	1·5 ,, 3	,,
Gum - - - -	2 ,, 3	,,
Lampblack - - - -	0·5 ,, 4	,,

The double picrates are obtained by mixing three equivalents of sodium picrate with one of lead or barium picrate.

Two varieties of Bronolithe have been approved for licensing, and are defined as follows :—

No. 1, consisting of a mixture of saltpetre, dextrine, sugar, and charcoal, with or without the addition of picrates of barium, sodium, and potassium, or any of them, and with or without the addition of nitronapthaline. The amount of picrates in the finished explosive is not to exceed 5 per cent. or the nitronapthaline 8 per cent.

* While this work was in the press a further sample of Britainite was examined, and it has now been put on the authorised list, the definition containing a proviso that the amount of chlorate of potash shall not exceed 1·3 per cent

Bronolithe—*continued.*

No. 2 is composed of the same ingredients as No. 1, but there is no limitation as to the amount of picrates. The total amount of saltpetre and nitro-napthalene is not to exceed 10 per cent.

114. **Brouillard.** *See* SIGNAUX.

115. I. **Brown Powder** is a special variety of gunpowder now largely used with heavy ordnance. Its manufacture originated in Germany, but is now carried on in this country. It differs from ordinary gunpowder in the proportions of its ingredients, which are for the description used in this country for military purposes:—

Saltpetre - - - - -	79 parts.
Sulphur - - - - -	3 „
Charcoal - - - - -	18 „

The charcoal used is made from straw, carbonised in a special manner, the details of which are kept secret. There are also some other peculiarities in the method of manufacture of this powder which cannot be given here. The finished powder is in the form of hexagonal prisms, compressed from grain powder, and having an axial perforation.

This powder gives very high velocities combined with moderate pressures on the gun. It is much less readily ignited than the ordinary powder, and burns without explosion (so far as is yet ascertained) in the open. Cartridges made up with it require a small primer of black powder to start the action.

The following comparison, made by Noble and Abel, is quoted in the "Journal of Society of Arts," 14.12.88:—

—	Cocoa Prism.	Black Pebble.
Units of heat evolved per gramme -	837	721
Volume of permanent gases in C. C. -	198	278

Other forms of brown powder have been recently introduced for sporting purposes. *See* PRISM BROWN, E.X.E., S.B.C.

116. III $_2$. **Brugère's Powder** consists of 54 parts of picrate of ammonia, and 46 of saltpetre. It is stable, safe to manufacture and handle, but rather expensive. It gives good results in a Chassepôt rifle. There is little smoke, and the residue is small, and consists of carbonate of potash. (D., p. 740.)

117. **Brunner.** *See* DYNAMOITE.

118. **Buchel.** *See* CARBONITE.

119. **Buchholz.** *See* CRAMER.

120. **Buckley.** *See* HARRISON.

121. **Budenberg.** *See* SHÄFFER.

121a. III $_1$. **Buechert** proposes a mixture of chloride or sulphate of ammonia, nitrate of soda, wood pulp, and nitro-glycerine. To prevent the action of the ammonia salt on the nitrate of soda, the particles of the former are coated with some protecting substance, such as oleate of alumina.

122. III $_1$. **Burstenbender** impregnates soft, spongy, elastic vegetable substances, such as cellulose, pith, fungi, &c., with glycocol or chondrin, and then mixes them with 20 to 60 per cent. of nitro-glycerine. The advantages claimed are non-liability to exudation up to 200° F. and freedom from freezing at temperatures below zero. It is stated that these properties are due to the glycocol or chondrin present. The explosive is granulated through sieves.

Chondrin is obtained from animal substances, and is very closely allied to gelatine. Glycocol is obtained by the action of acids on gelatine, and is a crystalline organic base. (T., p. 105.)

123. III $_1$. **Burton** has patented the combination of gunpowder with gun-cotton, with or without the addition of nitro-glycerine, nitro-gelatine, and gum-lac in solution. (Fr. Spec. 192,819, 6.9.88.)

124. III $_2$. **Cadoret** has proposed under the name of "tribenite" a powder consisting of picrate of ammonia, saltpetre, sulphur, charcoal, nitrate of ammonia, nitro-naphthaline, bichromate of ammonia, sugar, liquid hydrocarbons, &c. (P. & S. 149.)

125. **Caerphilly.** *See* NITROMAGNITE.

126. **Cahuc.** *See* SAFETY BLASTING POWDER.

127. III$_1$. **California Explosives Co.** propose a mixture of methyl and ethyl nitrates, nitro-benzol, methyl alcohol, pyroxylin, and nitro-glycerine, which has been specially purified by treatment with ethyl alcohol. (Spec. No. 11,326A, 3.7.91.)

128. III$_2$. **California Gun-cotton,** a mixture of 93 per cent. insoluble with 7 per cent. soluble nitro-cottons. (P. & S. 154.)

129. **Callow.** *See* MELVILLE.

130. **A.** III$_1$. **Camphorated Gelatine** is a special mixture of blasting gelatine and camphor.

131. **A.** III$_2$. **Cannonite.** Two varieties of this explosive have been licensed. Their definition is as follows:—

No. 1. Gun-cotton mixed or impregnated with a nitrate or nitrates (other than nitrate of lead and nitrate of ammonia) and resin, and with or without the addition of graphite.

No. 2. Gun-cotton mixed or impregnated with resin, and with or without the addition of graphite. (Spec. No. 1,115, 21.1.89.)

A sample of No. 2 gave:—

Nitro-cellulose - - -	86·32 per cent.
Resin - - - - -	11·28 ,,
Matter soluble in water - -	2·40 ,,

132. V$_1$. **Canouil's** cap composition consisted of:—

Chlorate of potash - - -	100 parts.
Powdered glass - - - -	100 ,,
Hydro-sulphite and cyanoferrique of lead - - - - -	80 ,,
Amorphous phosphorus - - -	2 ,,
Water - - - - -	200 ,,

The ingredients to be separately powdered, and mixed with the water to form a paste. (Spec. No. 970, 18.4.60.)

133. **A.** V_1. **Cap Composition.** The composition for percussion caps varies a good deal with the quality and object of the cap. That used for English Government black powder cartridges consists of:—

Fulminate of mercury	- - -	6 parts.
Chlorate of potash	- - - -	6 „
Sulphide of antimony	- - -	4 „

and in some special caps two parts of ground glass are added to the above. Some cheap caps contain no fulminate.

For smokeless powders a cap composition giving a larger flash with less detonating action, is desirable, and various mixtures are used for this purpose. For cordite the English Government use the following composition:—

Sulphide of antimony	- -	18 parts.
Chlorate of potash	- - -	14 „
Sulphur -	- - - -	1 „
Mealed powder	- - -	1 „
Fulminate of mercury	- -	6 „

134. **Carbo-azotine.** *See* SAFETY BLASTING POWDER.

135. **A.** III_1. **Carbo-dynamite,** patented by Mr. W. D. Borland, consists of:—

—	No. 1.	No. 2.
Nitro glycerine - - - - -	90 per cent.	50 per cent.
Cork charcoal - - - -	10 „	6 „
Saltpetre - - - - -	—	14 „

To every 100 parts of the explosive 1½ parts of carbonate of sodium or of ammonium may be added. In one variety water is added with the view of rendering the dynamite uninflammable or nearly so. This dynamite does not disintegrate or exude when exposed to the action of water. It is a black somewhat friable substance. The patentee claims that by the addition of 3 to 5 parts of his carbonised cork (in lieu of an equal amount of kieselguhr) to ordinary dynamite, the latter also becomes capable of resisting the action of water. (Spec. No. 758, 18.1.86. Fr. Spec. 188,007, 6.1.88.)

136. **A.** III_1. **Carbonite,** made by Bichel and Schmidt, of Schlebusch, is claimed to be a safe explosive for use in

Carbonite—*continued.*

fiery mines. Two varieties of it were submitted for licensing in this country in 1888, intended for use in coal and stone respectively. That for coal was a brown, friable mass; that for stone a black, moist, plastic one. They contained much the same ingredients, but in different proportions. The "stone" variety was rejected for exudation, the "coal" one was accepted, and is defined as consisting of not more than 27 parts of thoroughly purified nitro-glycerine (with or without the addition of not more than half a part of sulphuretted benzol) uniformly mixed with not less than 73 parts by weight of a pulverised preparation consisting of wood meal not less than 40 parts, nitrates of potassium and barium (or either of them) not more than 36 parts, and carbonate of sodium and lime (or either of them) not more than half a part, such preparation to be sufficiently absorbent when mixed in the above proportions to prevent exudation of nitro-glycerine. (Spec. No. 14,623, 11.11.86.)

137. **Caro.** *See* ANILINE FULMINANTE and CHROMATE DE BENZINE.

138. V_1. **Castan** has patented under the name of Vegetable Powder a mixture of :—

Chlorate of potash	- -	20 per cent.
Nitrate of potash	- -	48 ,,
Flowers of sulphur	- -	20 ,,
Wood meal	- - -	12 ,,

These ingredients are passed through a copper sieve in such a manner as to thoroughly mix them together. (Fr. Spec. No. 163,375, 18.7.84.)

139. III_1. **Castellanos Powders** are of two sorts. The first consists of nitro-glycerine with the addition of nitro-benzol, fibrous material, and pulverised earth. It is claimed that the addition of the nitro-benzol allows the nitro-glycerine to burn easily and rapidly without explosion and renders it less liable to freeze.

The second powder consists of nitro-glycerine, with nitrate of potash or soda, a picrate, sulphur, and a salt insoluble and incombustible in nitro-glycerine and carbon.

Castellanos Powders—*continued.*

The incombustible salts may be silicates of zinc, magnesia, and lime, oxalate of lime, carbonate of zinc, &c. The object of the inert salt is to render the nitro-glycerine less sensitive and safe. The following proportion is suggested:—

Nitro-glycerine - - - -	40 per cent.
Nitrate of potash, or soda - - -	25 „
Picrate - - - - -	10 „
Sulphur - - - -	5 „
Insoluble salt (as above) - - -	10 „
Carbon - - - - -	10 „

140. **Casthelaz.** *See* DESIGNOLLE.

141. V_1. **Castro Powder** consists of:—

Chlorate of potash - - - -	8 parts.
Bran - - - - -	7 „
Sulphide of antimony - - -	1 „

It is an American powder, proposed in 1884, and used in the form of cartridges.

142. **Catactines.** *See* CHANDELON.

143. **Cauvet.** *See* BARON and CAUVET.

144. **Celluloïdine.** *See* TURPIN (979).

145. **Celluloïque.** *See* TURPIN (979).

146. **Cellulosa.** *See* BJORKMANN, C. G.

147. III_1. **Chabert** has patented under the name of "Woodnite," a dynamite consisting of nitro-glycerine absorbed in woodmeal either simple or nitrated. (Fr. Spec. No. 191,906, 19.7.88.)

148. III_1. **Champion Powder,** an American dynamite identical or nearly so with Judson's Powder.

149. IV_2. **Chandelon** uses organic picrates (such as picrates of benzene, naphthaline, and of their nitro-derivates) in conjunction with nitrate of ammonium or other nitrates, or with chlorates partially or wholly replacing nitrates. (Spec. No. 13,360, 15.9.88.)

150. **Chanu.** *See* DAVY.

151. **Chapman.** *See* CANNONITE.

152. V_1. **Chapman** proposes to dispense with fulminates, and to obtain a longer flash and greater igniting power in cap composition. When used with high explosives there is supposed to be less risk of setting up detonation. One of the compositions is—

Amorphous phosphorus	15·90	per cent.
Potassium carbonate	2·00	"
Powdered resin	2·00	"
" cane sugar	2·00	"
" mercuric oxide	4·00	"
Peroxide of manganese	5·20	"
Magnesium	6·10	"
Potassium chlorate	10·90	"
" nitrate	51·90	"

(Spec. No. 16,997, 22.11.83.)

153. III_2. **Chemische Fabrik Griesheim** has patented the addition of a small proportion of trinitro-toluol or other similar body having a low melting point to picric acid for the purpose of binding together the crystals of the latter explosive. (Spec. No. 16,567, 17.6.93.)

153[a]. **Chili Saltpetre.** *See* NITRATE OF SODA, p. xxii.

154. **A.** III_2. **Chilworth Smokeless (Sporting) Powder** is defined as "consisting of thoroughly purified nitro-"cellulose, gelatinised by a suitable process, with or "without the addition of nitrates (other than nitrate of "ammonium.")

155. **A. Chilworth Special Powder.** *See* AMIDE POWDER.

156. **Chlorate of Potash** ($KClO_3$). *See* p. xxiii.

157. V_2. **Chloride of Nitrogen** ($N.Cl_3$.) is an oily liquid, highly explosive, and very sensitive to friction or shock. It is obtained by the action of chlorine on chloride of ammonium.

158. V_2. **Chromate de Benzine Diazotée** ($C_6H_5N_2 . HCrO_4$) is one of a general class of fulminating substances proposed by Caro and Griess. Amido-compounds are separated from their solutions by treating them with a misture of hydrochloric and chromic acids, or with the

Chromate de Benzine Diazotée—*continued.*

latter only, so as to obtain "a crystallised precipitate con-" sisting of a compound of hydrochloric acid and chromic " acid, with the nitrogenised bodies, or of compounds of " the latter with chromic acid only."

The process is stated to be the mixing of one equivalent of hydrochloride of aniline with two equivalents of hydrochloric acid. The product is treated with nitrite of lime, and the chromate is precipitated by adding a mixture of one equivalent of acid chromate of potash with one equivalent of hydrochloric acid. (Spec. No. 1,956, 28.7.66, and D., p. 742.)

159. **Chromated Gun-cotton.** *See* DAVEY (197).

160. **A.** III_2. **C. L. Powder** consists almost entirely of nitro-cellulose.

161. IV_2. **Clarite** is the name given to an explosive submitted for license in the Colony of Victoria and rejected. Three varieties were examined:—

—	No. 1.	No. 2.	No. 3.
	Per Cent.	Per Cent.	Per Cent.
Chlorate of potash	50	40	57
Xanthorœa balsam	50	30	29
Manganese dioxide	—	30	—
Camphor	—	—	14

162. III_2. IV_2. **Clark** proposed a wood gunpowder manufactured by treating grains of wood or wood pulp impregnated with alum or tannin with the usual acids. After nitration the grains are to be steeped in alum solution or boiled in potash solution, and then impregnated with a solution of nitrate of potash or soda. They may finally be coated with collodion. (Spec. No. 1,210, 11.4.68.)

He subsequently proposed to dilute the acid adhering to the pyroxiline after nitration with water "sufficient to " prevent oxidation of the pyroxiline" and to add to the moist compound "carbonate or bi-carbonate or chlorate " (!) or chromate of potash." A process admirably adapted to secure immediate explosion, or at least firing of the acid pyroxiline. Spec. No. 3,408, 10.11.68.)

163. III $_1$. **Clark's Explosive** is called in the specification glycero-pyroxiline, and is produced by treating textile or other vegetable fibre impregnated with glycerine with the usual acids. The idea is to get a compound of a more homogeneous character than a mixture of nitro-glycerine with pyroxiline. (Spec. No. 3,408, 10.11.68.)

164. **Clement.** *See* FUCHS.

165. III $_1$. **Coad's Explosive** is a mixture of nitro-glycerine with saltpetre and naturally decayed wood. There are several grades, of which the three following are examples:—

—	No. 1.	No 2.	No. 3.
	Per Cent.	Per Cent.	Per Cent.
Nitro-glycerine - - - - -	75	30	30
Saltpetre - - - - - -	5	50	—
Naturally decayed wood - - -	20	20	10
Ordinary blasting powder - -	—	—	60

According to Mr. Guttmann this explosive is identical with Rhexite. (T. 104.)

166. **Cock** has patented the coating of explosives with melted sulphur. (Fr. Spec. No. 165,127, 31.10.84.)

167. **Cocoa Powder.** *See* BROWN POWDER.

168. **A.** VI $_3$. **Colliery Safety Lighters** are an ingenious little arrangement for igniting safety fuze. The idea is the crushing of a tiny glass bead or tube containing sulphuric acid and embedded in a chlorate mixture. The acid ignites the latter, which in turn ignites the fuze, on to which the "lighter" contained in a metallic tube fits. Pinching the tube breaks the glass, and releases the acid. The object is to prevent the emission of flame or sparks when using the fuze in fiery mines. They are a revival of the old "Promethean match" with modifications Roth and also Zschokke have proposed the same idea.

169. **Collodine.** *See* VOLKMAN.

170. **A. III$_2$. Collodion Cotton,** nitro-cotton soluble in a mixture of ether and alcohol. *See* p. xxxv. For licensing purposes in this country, this explosive has been defined as consisting of thoroughly purified nitro-cotton, (*a*) of which not less than 15 per cent is soluble in ether alcohol, and (*b*) which contains not more than 12·3 per cent. of nitrogen. This definition is of course arbitrary, and would include varieties of nitro-cotton not suitable for conversion into collodion.

171. **Colloxyline.** *See* COLLODION COTTON.

172. **Cologne Powder.** *See* COLONIA POWDER.

173. III$_1$. **Colonia Powder** consists of a mixture of common blasting powder with 30 to 40 per cent. of nitro-glycerine. It has been manufactured at Cologne by Wasserfuhr Bros. (D., p. 721, T., p. 88.)

174. V$_1$. **Columbia Powder Manufacturing Co.** have patented a process of mixing chlorates and sulphur together with a nitrate and of forming an explosive in which the particles of chlorate and sulphur are coated with paraffin. (Fr. Spec. No. 217,632, 24.11.91.)

175. VI$_1$. **Combustible Cord** (A. Quentin) consists of a paste of nitro-glycerine, meal powder, and glycerine, intimately mixed together and made into fuze in paper tubes with india-rubber solution. The amount of glycerine regulates the rate of burning. (Spec. No. 1,805, 6.5.78.)

176. **A. III$_2$. Compressed Securite** has been licensed as consisting of a mixture of nitrate of potassium and nitrate of barium with thoroughly purified nitro-cellulose and one or more of the following substances thoroughly purified :—Meta-dinitro-benzol, dinitro-toluol, nitro-naphthaline, dinitro-naphthaline. A sample gave ;—

Meta-dinitro-benzol - - -	70·45 per cent.
Nitrate of potash - - - } Nitrate of baryta - - }	18·90 „
Nitro-cellulose - - -	10·65 „

177. **A.** III $_2$. **Cooppal's Powder** is practically much the same as Schultze gunpowder, and consists of nitro-lignin carefully purified, with or without an admixture of a nitrate or nitrates (other than nitrate of lead) and starch. Analysis of a sample gave :—

Nitro-cellulose	- - -	71·25 per cent.
Nitrate of barium	- -	23·65 ,,
Resinous matter	- - -	3·45 ,,
Moisture	- - -	1·62 ,,

178. **A.** III $_1$. **Cordite** is the smokeless powder adopted by the British Government. Its composition is as follows :—

Nitro-glycerine	- - - -	58 per cent.
Gun-cotton (insoluble)	- -	37 ,,
Mineral jelly (vaseline)	- -	5 ,,

The ingredients are kneaded together with acetone as a solvent, and the dough is then squirted through dies of various sizes according to the calibre of gun for which the Cordite is to be used. The cords so formed are then cut into lengths and the solvent is driven off. Finished Cordite is of a leathery consistency, and is yellowish brown in colour.

179. IV $_2$. **Cornet Powder** consists of 75 parts chlorate of potash to 25 parts resin. (O. G.)

180. **Corteso.** *See* MENDOCA.

181. IV $_2$. **Cotter's Powder** contains about equal parts of chlorate of potash and realgar.

182. **Cotton Powder.** *See* TONITE.

183. II. **Courteille's Powder** (also called Triumph Safety Powder) consists of the following ingredients :—

Nitrate of soda or potash	-	60 to 75 parts.
Sulphur - - - -	-	10 ,, 12 ,,
Charcoal - -	-	7 ,, 10 ,,
Peat and hard coal -	-	9 ,, 12 ,,
"Combined metallic sulphates"		2 ,, 4 ,,
Oleaginous matter, animal or vegetable (or tar), refined or crude - - - - -		1 ,, 3 ,,

The mixture is saturated with steam, and then heated by superheated steam till nearly dry, the temperature

Courteille's Powder—*continued.*

being gradually reduced from 250° F. to 150° F. The powder is then dried on hot plates. The advantages claimed are slow combustion, due to the use of peat, charcoal, and hard coal, also freedom from explosion in the open, or by friction or percussion. An essential principle claimed is that a comparatively large volume of the ingredients of gunpowder being mixed with a small volume of the other ingredients will, under proper conditions, when exploded in a close chamber or under pressure, form nitro-glycerine (!), or an equivalent thereof, and thereby constitute a powerful explosive. This claim appears to be founded on highly doubtful grounds, if indeed on any. (Spec. No. 3,217, 14.9.75, and T., p. 103.)

184. A. VII_2. **Crack Shots** are toy fireworks, consisting of amorces in which the charge is fulminate of silver, gummed on to a piece of paper having a strip impregnated with nitre.

185. II. **Craig** proposed to use nitrates of lime and magnesia in conjunction with nitrate of soda, and to coat each grain of powder with a film of collodion. (D., p. 605.)

186. I. **Cramer and Buchholz Powders,** the name given to various types of brown prismatic powder used in Italy.

187. **Cresilite.** *See* MELINITE.

188. III_2. **Cross, Bevan and Beadle** have patented a process of producing nitro-cellulose by first reducing the cellulose to hydro or oxy-cellulose by the action of acids or oxidants, and then dissolving in nitric acid. The solution is then precipitated with water or sulphuric acid according to whether lower or higher nitrates are required. (Spec. No. 9,284, 1.4.93.) *See* HYDRO-NITRO-CELLULOSE.

189. **Curtis and Andre.** *See* AMBERITE.

190. **Cycene.** *See* KITCHEN.

191. **Daddow.** *See* MINERS' SQUIBS.

192. **Dahmen.** *See* SAFETY DYNAMITE and BRITAINITE.

193. II. **Dahmenite** is licensed in the Colony of Victoria, as consisting of nitrate of potash, nitrate of ammonia, and naphthaline. (Spec. 23,579, 7.12.93.)

194. **Dale.** *See* ROBERTS.

195. III_2. **Darapski** proposed a powder very similar to Schultze. The name given to this was "Yellow Powder." (D. 669.)

196. III_2. **Davey** treated "as much starch or dextrine, or gum, or flour, or sugar, as will be dissolved" by boiling with a mixture of one part of nitric to three parts sulphuric acid. Hence he obtained "an acidulated preparation" which he proposed to mix in the proportion of 4 to 6 per cent. with the ingredients of ordinary gunpowder, the sulphur being "lessened or replaced" by the hydrocarbon. It is to be hoped that this was never tried anywhere on a manufacturing scale, as the conditions are excellently calculated to produce accidents. (Spec. No. 2,072, 21.7.62.)

197. III_2. **Davey** also proposed in another patent to treat gun-cotton with a solution of chromates, sesqui-chromates, or bi-chromates, with or without the addition of nitrate of potash (or similar salt) and gum or hydrocarbon.

This chromated gun-cotton he proposed to waterproof and use it in a mining fuze in lieu of ordinary gunpowder. (Spec. No. 2,832, 25.7.77.)

198. II. **Davey's Powder.** In this powder flour, bran, starch, or other glutinous or starchy matter is employed to replace a portion of the charcoal ordinarily contained in gunpowder. Nitrate of soda may also be substituted for saltpetre, if waterproof envelopes are provided for the powder. The object of the invention is to reduce the amount of smoke produced on firing.

The proportions given are:—

	Parts.			Parts.
Saltpetre - - -	64	or	Nitrate of soda -	63
Sulphur - - -	16		Sulphur - -	15
Charcoal - -	12			
Flour, bran, or starch	8			

(Spec. No. 2,478, 5.11.58.)

199. IV_2. **Davey's Powders** consist of:—

—	1.	2.
Chlorate of potash - - - -	6 parts.	6 parts.
Nitrate „ „ - - - - -	5 „	3 „
Yellow prussiate of potash - -	2 „	4 „
Bichromate of potash - - - - -	2 „	— „
Sulphide of antimony - - -	5 „	3 „

These compounds are chiefly intended for use as fuzes. (Spec. No. 14,065, 15.4.52.)

200. II. **Davey and Watson** proposed to impregnate gunpowder with liquid or gaseous hydrocarbons or melted solid hydrocarbons. (Spec. No. 2,641, 29.7.74.)

201. V_1. **Davies Powder** consists of:—

Yellow prussiate of potash - . . -	4 parts.
Chloride* of potassium - -	8 „
Loaf sugar - - - - - -	2 „
Crystallised sugar - - -	2 „
Sulphur - - - - - - -	1 „

It was proposed as a powder for firearms. The ingredients were to be mixed together when fired, being previously finely powdered. (Spec. No. 824, 30.3.60.)

202. II. **Davey** proposed to replace saltpetre by nitrate of soda in blasting powder.

203. III_1. **Dean's Explosive** consists of:—

Powdered nitro-cellulose or nitro-dextrine - - - - -	10 parts.
Water - - -	2 to 3 „
Nitro-glycerine - - - -	100 „

The idea is to make the nitro-glycerine safer to handle and transport by forming a pasty mass. (Spec. No. 2,226, 21.5.81.)

204. IV_2. **De Custro** has patented a mixture consisting of bran or other suitable cellulose, trisulphide, or sulphide of antimony, and saturated solution of chlorate of potash. (Fr. Spec. 159,172, 14.12.83.)

* Probably chlorate is intended.

205. **Deissler.** *See* KUHNT.

206. **De Lom de Berg.** *See* MAGNIER.

207. **De Mercader.** *See* DE TERRÉ.

208. **A.** III_2. **Denaby Powder** consists of compressed securite (*q.v.*) with the addition of charcoal.

Analysis of a sample gave:—

Dinitro-benzol - - -	21·49 per cent.
Nitro-cotton and charcoal -	5·07 „
Nitrates of potassium and barium.	73·18 „
Moisture - - -	0·26 „

209. III_2. **Designolle's Powders** were made in several varieties at Bouchon in 1869; they consisted of:—

—	For Torpedos and Shells.	For Guns.		For Small Arms.
		Ordinary.	Heavy.	
	Per Cent.	Per Cent.	Per Cent.	Per Cent.
Picrate of potash	55 to 50	16·4 to 9·6	9	28·6 to 22·9
Saltpetre - -	45 to 50	74·4 to 79·7	80	65·0 to 69·4
Charcoal -	— —	9·2 to 10·7	11	6·4 to 7·7

These powders were made much like ordinary gunpowder, 6 to 14 per cent. of moisture being added when being milled. The advantages claimed over gunpowder are greater strength, and consequently greater ballistic or disruptive effect, comparative absence of smoke, and freedom from injurious action on the bores of guns, owing to the absence of sulphur. (D., p. 738.) (Spec. No. 3,469, 5.12.67.)

210. IV_2, V_1. **Designolle and Casthelaz** proposed a variety of detonating powders consisting of two classes, one for general purposes, the other for "fulminating primings." In the latter class the oxidising effects of chlorate of

Designolle and Casthelaz—*continued.*

potash are stated to be overcome by the addition of salts of lead. The following proportions are given:—

—	Class I.		Class II.				
	1.	2.	1.	2.	3.	4.	5.
	Per Cent.		Per Cent.				
Picrate of potash - -	55	35	37	20	—	—	—
Chlorate of potash - - - -	47	47	18	18	16	26	50
Iso-purpurate* of potash -	—	—	—	—	—	—	50
Ferro-cyanide of potassium -	—	18	—	—	—	—	—
Chromate of lead - -	—	—	45	49	41	35	—
Picrate of lead - -	—	—	—	—	43	37	—
Charcoal - - - -	—	—	—	3	—	2	—

* $C_8H_4KN_5O_6$.

(Spec. No. 3,469, 5.12.67.)

211. II. **De Terré's Powders** consist of sawdust or similar carbonaceous substance, saltpetre, nitrate of soda, coal dust, lignite dust. He recommends the following proportions:—

For Blasting in Marble, Granite, or other Hardstone.		For Blasting in Limestone, Chalk, Coal, or other soft Stone.	
Sawdust - -	12·5 parts	Sawdust - -	11 parts.
Saltpetre - -	67·5 „	Saltpetre - - -	51·5 „
Sulphur - - -	20·0 „	Nitrate of soda - -	16 „
		Coal or lignite dust -	1·5 „
		Sulphur - -	20 „

(Spec. No. 2,715, 13.10.71.)

212. VI $_2$. **Detonating Fuze.** The French Explosives Committee have established a type of fuze consisting of tubes of lead or tin filled with gun-cotton and reduced to small diameter by drawing. (P. & S. 208.) *See* MAISSIN.

213. A. VI $_3$. **Detonators** are, as commercially used, metallic capsules, generally of copper, and resemble very long percussion caps. There are eight sizes usually made, which vary in dimensions and in the amount of explosive contained in them. The explosive is pure fulminate of mercury, or a mixture of the same with chlorate of potash and occasionally other substances. They are further distinguished as singles, doubles, trebles, &c., according to their number:—

No.	Explosive contained in each.	Weight of Explosives per 1,000.	Gross Weight per 1,000.
	Grs.	Lbs.	Lbs.
1 - -	4·6	0·66	2·1
2 - - -	6·2	0·88	2·8
3 - -	8·3	1·19	3·3
4 - - -	10·0	1·43	3·7
5 - -	12·3	1·76	4·1
6 - - -	15·4	2·20	5·0
7 - -	23·1	3·31	?
8 - - -	30·9	4·41	11·7

The above figures vary somewhat with different makers.

214. II. **De Tret's Powder,** known also as Pyronome, consists of:—

Nitrate of soda - - - -	52·5	per cent.
Spent tan - - -	27·5	,,
Sulphur - - - - -	20·0	,,

The ingredients are boiled together and dried. It is a slow-burning powder of no great strength. (Spec. No. 1,226, 17.5.59.)

215. III $_2$. **Deutsche Sprengstoff Actien-Gesellschaft** (Die) have patented the manufacture of nitro-cellulose capable of being converted into a fine dense grain, from the shells of nuts and fruit-stones rich in cellulose. (Fr. Spec. No. 172,309, 16.11.85.)

216. III $_1$, **Diaspon** (Johann Anders) consists of:—

Nitro-glycerine - - -	47 to 63	parts.
Nitro-cotton - - -	0·5 to 3·0	,,
Nitrate of soda - - -	22 to 23	,,
Wood cellulose - - -	8 to 18	,,
Sulphur - - - - -	3 to 9	,,

(Spec. No. 81, 24.2.81.)

217. III_1. **Diaspon-Gelatin** (Johann Anders) consists of 92 to 95 parts of nitro-glycerine mixed with 5 to 7 parts of "slightly nitrated" wood cellulose or collodion wood, and 0·5 to 2·0 parts of alcohol. The ingredients are placed in a water bath at 40° to 45° C. with 10 to 15 parts of a solvent composed of 6 parts of ether to 2 parts of alcohol. The nitro-cellulose is dissolved, and the ether, with the added alcohol, evaporates. (Spec. No. 80, 24.2.81.)

218. III_2. **Dieckerhoff's Powder** (*see also* HERAKLIN) consists of gunpowder mixed with not more than 15 per cent. of precipitated alkaline picrate or picrates, the object being to increase the explosive power of gunpowder without increasing its liability to accident or ignition. It is claimed also that a mixture of nitrate with sulphur and picrate in the above general proportions, without charcoal, is practically equivalent to the above, as experience has shown that charcoal is not essential to the compound. (T., p. 106.)

219. **Diessler.** *See* KUHNT, WETTER.

220. **A.** III_2. **Di-Flamyr,** signifying in Welsh flameless, consists of gun-cotton mixed with nitrates.

221. III_1. **Diller's White Dynamite** is of the following composition:—

Nitro-glycerine - - -	70 per cent.
Guhr calcaire - - -	19·35 ,,
Wood pulp - - - -	10·65 ,,

The guhr calcaire is a calcareous tuff found in stalactite caves.

222. **Dinitro-benzol,-Naphthaline, &c.** *See* NITRO-BENZOL, &c.

223. III_1. **Dinitro-glycol** is a liquid much resembling nitrate of methyl, *q.v.* Glycol is ethylene alcohol $C_2H_4(OH)_2$ and the dinitro-glycol is $C_2H_4(O.NO_2)_2$ and hence is a nitric ester.

224. III_2. **Diorrexin,** an Austrian explosive, consisting of:—

—	A.	B.
	Per cent.	
Nitrates of potash and soda - - - - - - -	75	60
Sulphur - - - - - -	12	12
Sawdust - - - - - - - - - -	13	10
Charcoal - - - - - - - -	—	7
Picric acid - - - - - - - - -	—	1·5
Moisture - - - - - - -	—	7·5

("Mining Journal," 18.6.87.)

225. **R.** IV_2. **Diripsite.** Two samples of an explosive bearing this name were submitted in 1889 by Mr. H. C. Williams. The first was rejected for extreme sensibility to friction and percussion, the second for want of chemical stability. It was a chlorate explosive, but the inventor desired its composition to be considered confidential.

226. **A.** VII_2. **Distress Signal Rocket** consists of a sound signal rocket (*q.v.*), with the addition of coloured stars.

227. III_1. **Dittman** proposes to render the transport and storage of nitro-glycerine safer by mixing it with a porous combustible substance, such as finely divided wood charcoal saturated with a solution of saltpetre or nitrate of soda and with a solution of carbonate of soda. Or the nitro-glycerine may be mixed with nitro-cellulose, or sawdust, or nitrated woody matter. (Spec. No. 3,458, 5.12.67.)

228. **Dittmar.** *See* DUALINE, GLUKODINE, TITAN, and XYLOGLODINE.

229. **Divine's Explosive.** *See* RACK-A-ROCK.

230. V_1. **Domergue's Explosive** is a rough mixture of chlorate of potash and sulphur.

231. **Doutrelepont.** *See* PETRAGITE.

232. I. **Drayson** proposed to dissolve saltpetre in half its weight of warm water, and to mix the solution with the requisite proportions of sulphur and charcoal in the mill. The ingredients thus mixed were to be ground, dried, and finished in the usual manner. (Spec. No. 2,427, 31.10.55.)

In a subsequent specification he proposed to inject steam into a mixture of the usual ingredients enclosed in a vessel, and in quantities sufficient to give adhesiveness to the composition. He proposed further to grain the powder direct from mill cake instead of first pressing it in the usual way. (Spec. No. 292, 4.2.64.)

Drayson's process was tried at the Dartmoor Gunpowder Mills, but abandoned, chiefly in consequence of a serious explosion.

233. III 1. **Dualine** consists of:—

Nitro-glycerine - - - - -	50 per cent.
Fine sawdust - - - -	30 „
Saltpetre - - - - - - -	20 „

but various mixtures of the same kind are made under the same title; the sawdust is sometimes nitrated. Dittmar's patent speaks of a mixture of cellulose, nitro-cellulose, nitro-starch, nitro-mannite, and nitro-glycerine. The cellulose is prepared from soft woods, treated with dilute acids, and then boiled in a solution of soda.

Dualine is said to be liable to exudation. The commercial variety is a yellowish brown powder. It is said to be more sensitive than ordinary dynamite to heat and friction. Owing to the excess of carbon the gases evolved on explosion contain a large proportion of the poisonous carbonic oxide. (D., p. 726. T., pp. 87 and 100. Spec. No. 3,088, 3.9.75.)

Another variety was patented by Schultze (Spec. 2,542, 14.8.68), and consists of a mixture of from 10 to 60 lbs. weight of nitro-glycerine with 100 lbs. of wood gunpowder, the proportion being varied according to the purposes for which the explosive is required. It is also prepared with an admixture of small grains of wood or powdered charcoal impregnated with a nitrate.

234. IV_1. **Dulitz** makes a jelly of gun-cotton in nitro-benzol, and adds this to four times its weight of chlorate of potash. Up to 10 per cent. of another oxidising agent may be substituted for an equivalent amount of the chlorate. (Spec. No. 12,838, 17.8.86.) *See* also KINETITE.

235. **Dumas.** *See* NITRAMIDINE.

236. **Duplexite.** *See* TURPIN (980).

237. II. **Dupont's Brown Powder** consists of:—

Saltpetre	78·0 parts.
Sulphur	2·8 to 3·0 ,,
A carbohydrate (*e.g.*, sugar)	3·0 to 4·0 ,,
Baked wood	14·0 to 12·5 ,,

(A. and E. I., 214.)

238. III_1. **Dupont's Smokeless Powder** consists of gun-cotton gelatinised and incorporated with a deadening agent such as nitro-benzol, nitro-naphthaline, camphor, &c. (Spec. No. 15,865, 14.10.93.)

239. I. **Durnford** uses light inflammable charcoal made from cork in the proportion of 20 parts charcoal to 80 parts of saltpetre, with or without the addition of 1 to 10 per cent. of sulphur. He claims this as "smokeless" and non-hygroscopic powder. (Spec. No. 3,578, 13.3.86.)

240. **Durnford** has patented a method of removing the water from nitro-cellulose by treating the wet material with a spirit soluble in water. The object of the invention is to do away with the necessity of drying gun-cotton by heat when it is required to treat it subsequently with a solvent. (Spec. No. 20,880, 17.11.92.)

241. III_2. **Duttenhofer Powder,** a smokeless powder, which was adopted by the German Government and afterwards discarded. The exact composition is kept secret, but it is believed to be covered by the patents of Glaser and Engel, in which case it probably consists of gun-cotton dissolved in a solvent and mixed with a nitrate and a small quantity of a hydrocarbon, such as paraffin. The powder is in the form of small hard grains irregular in shape but uniform in size. It is black in colour, having been probably treated with graphite. (Specs. No. 17,167, 13.12.87, and 6,022, 25.4.87.)

242. **Dynamagnite.** *See* NITRO-MAGNITE.

243. **A.** III$_1$. **Dynamite No. 0** is defined as consisting of not more than 75 per cent. of nitro-glycerine mixed with 25 per cent. of woodmeal, sufficiently absorbent in quality to prevent exudation. Carbonates of lime and magnesia may be added to the extent of 1½ per cent.

244. **A**. III$_1$. **Dynamite No. 1.** This is the ordinary dynamite of commerce as used in this country. It is an admixture of nitro-glycerine with a porous infusorial earth called *kieselguhr*, which consists mainly of silica. Beds of it are found in Germany, Scotland, and elsewhere. It is calcined to drive off water and organic matter, and more or less of the accompanying sand is sifted away. It generally contains a little iron, which accounts for the more or less red tinge observable in ordinary dynamite. A little carbonate of ammonia or soda is usually added to the dynamite, so that its aqueous extract is alkaline.

Till 1887 this explosive was licensed in the following terms:—

"*Dynamite No.* 1, consisting of not more than 75 parts by weight of thoroughly purified nitro-glycerine, uniformly mixed with 25 parts by weight of an infusorial earth known as *kieselguhr*, and sufficiently absorbent in quality when mixed in the above proportions to prevent exudation of nitro-glycerine."

The above definition is now slightly altered, so as to include not only a simple mixture of nitro-glycerine and *kieselguhr* but also a mixture containing small quantities of other substances. The definition now runs thus:—

"*Dynamite No.* 1, consisting of not more than 75 parts by weight of thoroughly purified nitro-glycerine, uniformly mixed with 25 parts by weight of:—

"(*a*.) An infusorial earth known as *kieselguhr*, or,

"(*b*.) A non explosive mixture of *kieselguhr* with such other ingredients and in such proportions as may

Dynamite No. 1—*continued.*

for the time being be sanctioned by the Secretary of State;

Provided:—

"(1.) That the said (*a*) *kieselguhr, or* (*b*) non-explosive mixture, shall be sufficiently absorbent in quality when mixed in the above proportions to prevent exudation of nitro-glycerine; and

"(2.) That there may be added to the *kieselguhr* or non-explosive mixture an amount of carbonate of ammonia not exceeding $1\frac{1}{2}$ parts by weight in every 100 parts by weight of the finished dynamite."

The ingredients at present sanctioned under (*b*) are:—

Ingredient	
Carbonate of sodium	In all 8 parts (or less), by weight in substitution for an equal amount by weight of *kieselguhr.*
Sulphate of barium	
Mica	
Talc	
Ochre	

Provided that the total amount of carbonate of sodium present shall in no case exceed three parts by weight in every 100 parts by weight of the finished dynamite.

Dynamite is a plastic mass varying in colour from buff to reddish brown. The direct contact of water disintegrates it, separating the liquid nitro-glycerine, and hence great caution is requisite in using it in wet places. In this particular it compares unfavourably with blasting gelatine and gelatine dynamite. As a rule it freezes at about 40° F. and remains frozen at temperatures considerably exceeding that point. In this condition it is practically useless as a blasting agent, and requires to be thawed or "tempered." This operation requires great care, and the instructions issued with every packet should be closely and carefully attended to.

When ignited in comparatively small quantities it simply burns away fiercely, but fatal accidents have arisen in considerable number from persons supposing that, as it is reasonably safe to ignite a few cartridges of unfrozen dynamite, it is equally safe to warm it upon a shovel, in an oven, in a tin over a fire, or in various other ways which usually lead to a verdict of "Accidental Death."

Dynamite No. 1—*continued.*

It cannot be too strongly impressed that if dynamite or other nitro-glycerine preparations are gradually warmed up to a temperature approaching their exploding point (about 360° F.) they become extremely sensitive to the least shock or blow, and once that point is reached they do not simply ignite but explode with great violence.

Frozen dynamite has some peculiar properties as distinguished from unfrozen dynamite, which have been experimentally shown and may be summarised as follows:—

(1.) It is considerably less sensitive to a blow, or to the impact of a projectile. A rifle bullet fired into unfrozen dynamite readily explodes it, but a bullet from a Martini-Henry rifle fails to do so when the explosive is frozen, at 25 yards range. Blasting gelatine on the other hand is more sensitive to similar treatment when frozen.

(2.) It is much more susceptible to explosion by simple ignition. For instance, so small an amount as 1 lb. has been exploded by ignition when frozen, while very considerably larger quantities of the unfrozen material will burn away without explosion.

(3.) Like frozen nitro-glycerine, it is much less sensitive to detonation.

Dynamite is almost invariably issued in cylindrical rolls, of a diameter suitable to the bore-holes in which it is to be used, and these rolls are wrapped in parchment paper to form cartridges. Five pounds of cartridges make a packet, and ten packets go to a box, which thus contains 50 lbs. of dynamite.

Blasting gelatine and gelatine dynamite are issued in the same way.

245. **A. III$_1$. Dynamite No. 2** is milder and slower than No. 1, and was introduced to compete with gunpowder where the great power and local shattering effect of No. 1 was undesirable, for instance in coal mines, and slate and granite quarries. It is issued like No. 1, but is easily distinguished by its black colour. It is, however, little, if at all, now used in this country.

Dynamite No. 2—*continued.*

In the license it is defined as follows:—

"*Dynamite No.* 2, consisting of not more than 18 parts by weight of thoroughly purified nitro-glycerine, uniformly mixed with 82 parts by weight of a pulverised preparation, composed of nitrate of potassium 71 parts, charcoal not less than 10 parts, and purified paraffin (*or* ozokerit) 1 part (*or* nitrate of potassium 72 parts, and charcoal not less than 10 parts) by weight, and sufficiently absorbent in quality when mixed in the above proportions to prevent exudation of nitro-glycerine."

246. III. **Dynamite au Charbon,** proposed by Nobel, is of two kinds, to either of which sulphur may be added. The compositions are as follows:—

—	I.	II.
	Per Cent.	Per Cent.
Nitrete of baryta	68	70
Charcoal	12	—
Resin	—	10
Nitro-glycerine	20	20

It is practically much the same as Dynamite No. 2. (D., p. 720.)

247. III$_1$. **Dynamite blanche de Paulilles** is much the same as Dynamite No. 1, being composed of 70 to 75 per cent of nitro-glycerine mixed with 30 to 25 per cent. of a natural silicious earth. (D., p. 702.)

248. III$_1$. **Dynamite d'Ammoniaque.** *See* AMMONIA DYNAMITE.

249. III$_1$. **Dynamite de Boghead** consists of 60 to 62 per cent. of nitro-glycerine mixed with carefully cleansed and powdered ashes of Boghead mineral. The ash consists of a mixture of silica and alumina.

250. III. **Dynamite, E.C.,** is identical with the old definition of Dynamite No. 1, except that carbonate of soda (not exceeding 3 per cent.) is added.

251. **Dynamite Gomme.** *See* GELATINE DYNAMITE.

252. III$_1$. **Dynamites grises de Paulilles** are mixtures of 20 to 25 per cent. of nitro-glycerine with nitrate of soda and resin, with or without charcoal. They are therefore much the same as Dynamite au Charbon. (D., p. 721.)

253. III$_1$. **Dynamite de Krümmel** is a sort of dualine. The composition consists of:—

—	No. 1.	No. 2.
Nitro-glycerine - - -	48 to 50 parts	30 to 35 parts
Dried and nitrated sawdust -	10 "	60 "
Kieselguhr - - - -	40 "	5 "

They are brownish in colour. The weaker mixture is designed for use in coal and other soft materials. (D., p. 726.)

254. III$_1$. **Dynamite Noire** is a mixture of powdered coke and sand with about 45 per cent. of nitro-glycerine. It is said to be more dangerous than ordinary dynamite. (D., p. 719.)

255. **Dynamite Paille.** *See* STRAW DYNAMITE.

256. III$_1$. **Dynamite Rouge** consists of 66 to 68 per cent. of nitro-glycerine absorbed in tripoli, a very finely divided silica. (D., p. 702.)

257. III$_1$. **Dynamite, S. 1,** is identical with Dynamite, E.C.

258. III$_1$. **Dynamit de Trauzl.** *See* TRAUZL.

• 259. III$_1$. **Dynamites de Vonges** consists of :—

—	No. 1.	No. 2.	No. 3.	Special.
	Per Cent.	Per Cent.	Per Cent.	Per Cent.
Nitro-glycerine - -	75	50	30	90
Randanite * - -	20.8 }	—	—	1 }
Silice de Vierzon - -	3·8 } 25	48 }	—	— }
Sub-carbonate of magnesia.	0·4 }	— } 50	—	1 }
Meudon chalk - -	—	1·5 }	—	— }
Red ochre - - -	—	0·5 }	—	— } 10
Silice de Launois -	—	—	60 }	— }
Blast furnace slag - -	—	—	4 }	— }
Carbonate of lime -	—	—	1 } 70	— }
Yellow ochre - - -	—	—	5 }	— }
Silice spéciale - -	—	—	—	8 }

* A silicious material like kieselguhr found in the neighbourhood of Ceyssat (Puy-de-Dôme), and attributed to the decomposition of felspar by natural accidulated mineral water.

(D., p. 701.)

260. **Dynamite de Zamky.** *See* DYNAMITE DE KRÜMMEL.

261. III$_2$. **Dynamo** is licensed in the Colony of Victoria as consisting of perchlorate, nitrate, and picrate of potash, tar, and sawdust.

262. IV$_2$. **Dynamogen** consists of :—

Yellow prussiate of potash - -	17 parts.
Water - - - -	150 ,,
Charcoal - - -	17 ,,

These are boiled, well stirred, and allowed to cool. There are then added :—

Potash - - -	35 parts.
Chlorate of potash - - -	70 ,,
Starch - - - -	10 ,,
Water - - - - -	50 ,,

The whole is made into a thin paste and brushed on to filtering paper, three coats being laid on each side. The paper so impregnated is dried, cut up, and rolled into cartridges. (Spec. No. 2,895, 19.6.82.)

263. III$_1$. **Dynamo-electric,** a nitro-glycerine explosive for which extravagant advantages are claimed without any adequate proof. It was invented by Major Verstraete, of the Belgian army. It is apparently a crude form of dynamite with an active base.

264. IV$_2$. **Dynamoïte.** This explosive resembles asphaline and is composed as follows :—

—	A.	B.	C.
Malt coombes - -	Per Cent. 40 to 70	Per Cent. 30 to 70	Per Cent. 40 to 70
Nitrate of ammonia -	40 to 10	70 to 30	—
Chlorate of potash - -	20 to 40	—	60 to 30

(Spec. No. 5,843, 4.4.91.)

According to the French specification the proportions would be 60 per cent., 15 per cent., and 25 per cent. (Fr. Spec. 212,565, 4.4.91.)

265. VI$_1$. **Eales** has patented a blasting fuze of which the composition consists of black powder and gun-cotton. The latter ingredient may be rendered less rapid in combustion by treating it with a solution of nitrate or chromate of potash. (Fr. Spec. 184,326, 20.6.87.)

266. II. **Eaton's Powder,** a coarse-grained, slow-burning powder, containing nitrate of soda. It has been tried in firearms mixed with quicker powder, the object being to obtain a sort of progressive powder. (D., p. 607.)

267. III$_2$. **Eclipse Powder,** proposed as a "smokeless powder," consists of picric acid, litharge, sulphur, and charcoal.

268. A. **E.C. Dynamite.** *See* DYNAMITE (250).

269. A. **E.C. Powder Company's Rifle Powder.** *See* JOHNSON-BORLAND POWDER.

270. A. III$_2$. **E.C. Sporting Powder** is licensed as consisting of nitro-cellulose mixed with nitrates, with the addition of colouring matter (aurine or ultramarine), and with or without beeswax, paraffin, shellac, gums, or resins.

Two samples gave :—

—	I.	II.
	Per Cent.	Per Cent.
Nitro-cellulose soluble	27·95	21·79
„ „ insoluble	28·35	25·58
Cellulose (unconverted)	3·15	4·17
Nitrates of potassium and barium	37·80	38·32
Matters soluble in benzol	0·60	1·95
„ „ „ alcohol	2·15	6·32
Moisture	—	1·87

271. III$_2$. **Ecrasite,** an explosive said to have been adopted by the Austrian Government. It consists of trinitro-cresylate of ammonia and nitrate of potash, and therefore closely resembles picric powder. It is stemmed into shells or compressed into cartridges, and in this condition it has a density of about 1·4. A gun-cotton primer is generally used, but the mixture can be readily exploded with a 2 gramme detonator. It is somewhat hygroscopic.

272. **Edison.** *See* PREISENHAMMER.

272a. III_2, V_1. **Edmunds** has patented the following mixtures:—

—	I.	II.	III.
	Parts.	Parts.	Parts.
Chlorate of potash - - -	6	—	4
Chloride of potassium - -	4	4	—
Nitrate of potash - - -	12	12	4
Sulphur - - -	1	3	$1\frac{1}{3}$
Picric acid - - - -	3	3 to 4	$2\frac{1}{3}$
Prussian blue - - -	1	$1\frac{1}{3}$	$\frac{2}{3}$
Sawdust - - - -	5	8	—
„ (nitrated) - -	—	—	3
Silica - - - -	—	—	1

The mixtures may be impregnated with a solution of rubber. (Spec. 23,416, 5.12.93.)

273. IV_2. **Ehrhardt's Powders** consist of:—

—	For Artillery.	For Blasting.	For Shells.
Chlorate of potash -	1 part	1 part	1 part
Saltpetre - -	1 „	1 „	— „
Charcoal - - -	— „	4 „	— „
Tannin - -	1 „	2 „	1 „

(D., p. 613.)

He also proposed a mixture of about equal parts of tannic acid or resin, and chlorate or nitrate of potash finely powdered. (Spec. No. 1,694, 8.7.64.)

Also a mixture of the same ingredients in the proportion of one to three. The powders were to be placed in a flask with two chambers, and so kept separate till mixed in the act of loading. (Spec. No. 2,594, 20.10.64.)

Also a mixture of:—

—	For Blasting.	For Arms.
Chlorate of potash -	1·5 parts (by bulk) -	4 parts (by weight).
Nitrate of „ -	1·5 „ „ -	— „
Cutch or tannin - -	1·0 „ „ -	1 „
Cannel coal - -	2·0 „ „ -	— „

(Spec. No. 402, 13.2.65.)

274. II. **Eisler's Powder** consists of:—

Nitrate of soda - - - 70·1 parts.
Sulphur - - - 17·6 „

Eisler's Powder—*continued.*

Charcoal	-	12·25 parts.
Sugar	-	1·2 ,,
Ashes	-	1·2 ,,
Moisture	-	0·83 ,,

(O. G.)

275. A. **Electric Detonators.** *See* ELECTRIC DETONATOR FUZES.

276. A. VI$_3$. **Electric Detonator Fuzes** consist of electric fuzes as described below with detonators attached.

277. A. VI$_2$. **Electric Fuzes** are of two kinds, high tension and low tension. In the former the terminals are embedded in the composition, and in the latter they are connected together by a fine wire which is surrounded by the explosive material. The following compositions have been licensed:—

High Tension.

Name.	Composition.
Abel, &c.	Sulphide of copper, phosphide of copper, and chlorate of potash.
Brain, &c.	Sulphide of antimony, sub-phosphide of copper, and chlorate of potash. Sulphide of antimony, platinum, silver, and chlorate of potash.
——	Sulphide of antimony, silver precipitate, plumbago, and chlorate of potash.
Spon	Sulphide of antimony, chlorate of potash, with or without carbon.

Low Tension.

Name.	Composition.
Abel, &c.	Gunpowder and gun-cotton.
Smith	Fulminate of mercury (not more than ½ grain).
——	Gunpowder. Gun-cotton. Collodion cotton. Chlorate of potash and sulphide of antimony.
Spon	Gun-cotton, chlorate of potash, and powdered galls.

278. III$_1$. **Electric Powder,** an American dynamite with active base, containing from 25 per cent. to 33 per cent. of nitro-glycerine. (P. & S. 299.)

279. A. VI$_2$. **Electric Primers** have the same definition as low tension electric fuzes.

280. A. VI$_2$. **Electric Tubes** have a similar definition to electric fuzes. They are used for firing heavy guns. Two varieties are licensed, viz., Abel's and Elswick. The former contain the same composition as Abel's electric fuzes, and the latter contain gun-cotton.

280[a]. III $_2$. **Electronite** consists of blasting amberite with the addition of 40 per cent. of carbonate of ammonia. This explosive is now (December 1894) under examination with a view to its being placed on the authorised list.

281. A. **Elswick Electric Tubes.** *See* ELECTRIC TUBES.

282. A. VI $_3$. **Elswick Mechanical Tubes,** for firing guns, consist of cases of metal or other suitable material containing their own means of ignition, and a charge not exceeding one ounce of gunpowder.

282[a]. A. III $_2$. **Emerald Powder** consists of Cooppal's powder (*q.v.*), coloured with malachite green (oxalate of tetramethyl-diamido-triphenyl-carbinol).

283. III $_2$. **Emilite** (P. M. E. Audouin). The portion of coal tar boiling between 185° and 200° C. is nitrated, after removing resinous matters. (Spec. No. 5,899, 22.4.87.)

284. III $_2$. **Emmens** has patented a mixture of a nitrated hydrocarbon such as picric acid, with an alkaline nitrate such as nitrate of soda. The two ingredients are mixed in the fused condition. (American Spec. No. 422,514, 4.3.90.)

285. III $_2$. **Emmens** has also patented an explosive consisting of nitro-paper, impregnated with ammonia and picric acid. (American Spec. No. 423,230, 11.3.90.)

286. III $_2$. **Emmensite.** "Emmens" crystals are produced by treating picric acid with red fuming nitric acid (S. G., 1·52.) The acid dissolves with the evolution of red fumes, and when cooled* yields crystals stated to be different to picric acid. The liquor deposits a second crop of similar crystals and a quantity of lustrous flakes. These flakes, when heated in water, separate into two new bodies; one of these enters into solution and forms crystals unlike the first, while the other body remains undissolved.

The acid crystals and residuum are used in conjunction with a nitrate to form an explosive. (Spec. No. 370, 10.1.88.) "Journal of Society of Chemical Industry," December 1888.

---

* This is not very intelligible as the solution of picric acid in nitric acid produces a lowering of the temperatures.

Emmensite—*continued.*

Emmens' acid would have for formula:—

$$C_{13}H_8O.(NO_2)_6.2OH,$$

and is therefore an intermediate between trinitro-phenol and trinitro-cresol.

By a subsequent patent Emmens has proposed to lower the melting point of the above described acid crystals by the addition of some allied nitrated hydrocarbon such as nitro-benzol. (American Spec. No. 422,515, 4.3.90.)

287. **Engel.** *See* DUTTENHOFER.

288. III$_1$. **Engels** has proposed some rather complicated explosives consisting of:—

Pyroxiline - - -	5 to 10 per cent.
Nitro-glycerine - - -	60 to 70 ,, ,,
Pyro-papier - -	15·5 to 18 ,, ,,
Nitro-starch - - -	0·5 ,, ,,
Nitro-mannite - -	5 to 1 ,, ,,
Nitro-benzol - - -	0·5 ,, ,,
Ammonium salts (nitrate sulphate or chloride)	10 to 30 ,, ,,
Water glass - -	0·5 ,, ,,
Saltpetre - - - -	8 to 10 ,, ,,

They thus appear to be a jumble of all sorts of nitro-compounds with the addition of nitrates. (M. No. XIII., p. 246.)

289. **A.** II. **Espir's Powder** was licensed for manufacture in this country at a factory near Plymouth, but the manufacture has been abandoned, and the amount hitherto consumed has not reached any great proportions. The license defines it as a mechanical mixture of nitrate of soda, sulphur, and sawdust. The specification gives the following composition:—

Nitrate of soda - - -	60 per cent.
Sulphur - - - - - -	14 ,,
Sawdust - - - -	26 ,,

It was made in a wet state, one pint of boiling water being used to dissolve two pints of nitrate of soda. The solution was poured on to, and well mixed with, the other two ingredients, and the whole dried and sifted. Close confinement is necessary to develop the explosive forces of this powder, it being stated to be quite non-explosive in the open air. It is stated to improve by age. (Spec. No. 291, 26.1.75.)

290. **Esselens.** *See* WYNANTS.

291. **Etna Powder.** *See* ÆTNA POWDER.

292. R. V_2. **Etnite** practically consists of asphaline, with the addition of about 8 per cent. of sulphide of antimony.

293. I. **E.X.E. Powder** (experimental E), a prismatic powder in use in the service for medium guns. The prisms are 0″·9764 in height, and 1″·3662 in diameter (over the sides). E.X.E. differs from black and brown powders in the proportion of the ingredients and the composition of the charcoal. The latter is in fact a mixture of two charcoals, each of which has a distinctive character. The powder is of a dark slate colour, and each prism has on one face an indented ring round the centre perforation. The density is 1·8.

294. IV_2. **Explosive Paper** is much the same as Dynamogen and other similar explosives, consisting of rolls of paper impregnated with a chlorate mixture. The ingredients in this case are:—

Saltpetre	5 parts.
Chlorate of potash	5 ,,
Charcoal or Powdered coal	1 ,,
Fine hardwood sawdust	1 ,,

with a little gum for binding, and enough water to make a paste. (M., No. VII., p. 115.)

295. **Explosive-Resin.** *See* NITRESINE, PELLIER (721).

296. III_1. **Extra Dynamite** is made by some Continental factories, and consists of 48·4 parts of nitro-glycerine, 1·6 of nitro-cotton, 34·5 of nitrate of ammonia, 5 of nitrate of soda, 9 of rye flour, 1 of soda, and ½ of ochre.

297. **Extra Hercules Powder.** *See* DYNAMAGNITE.

298. IV_2. **Extralite** consists of:—

Nitrate of ammonia	50 parts.
Carbonate of ammonia	5 ,,
Liquid hydrocarbons	10 ,,
Solid hydrocarbons	5 ,,
Chlorate of zinc	50 ,,

(H. M.)

299. III $_1$. **Extra Powder** is a variety of giant powder from which it differs in containing a proportion of nitrate of ammonium. The hygroscopic qualities of this salt are neutralised by coating it with crude vaseline before mixing it with the other ingredients.

300. **Fahneljelm.** *See* SEBASTINE.

301. III $_2$. **Falkenstein and Böhm** propose a mixture consisting of nitro-cellulose steeped in permanganate of potash. The resulting product is afterwards converted either into a smokeless powder or into an explosive for blasting purposes. In the latter case bichromate of potash is added. (Spec. No. 7,238, 14.4.93.)

302. **Faille.** *See* GRISOUTITE, MELANITE.

303. IV $_1$. **Fallenstein** has patented a mixture consisting of persulphide of antimony, nitrates, chlorates, and solution of nitro-cellulose in nitro-benzol or other isomeric body. (Fr. Spec. No. 163,256, 23.10.86.)

304 **Faure.** *See* MACKIE.

305. III $_1$. **Favier's Explosive** consists of a compressed hollow cylinder composed of 91·5 parts of nitrate of ammonium and 8·5 parts of mono-nitro-naphthaline, filled inside with loose powder of the same composition. The cartridges are enclosed in waterproof wrappers. This explosive has been licensed in this country under the name of "Miners' Safety Explosive." In the specification the idea is started of filling the central cavity with a stronger explosive, *e.g.*, dynamite, gun-cotton, &c. (Spec. No. 2,139, 16.2.85.)

As sold in France this explosive consists of nitrate of ammonia with 8·5 per cent. to 12 per cent. of dinitro-naphthaline and 4·5 per cent. of trinitro-naphthaline. Formerly in that country an explosive was made under this name, consisting of 80 per cent. nitrate of soda, and 20 per cent. mono-nitro-naphthaline. (P. & S. 322.)

306. **Favierite.** *See* FAVIER.

307. **Fehleisen.** *See* BLECKMANN.

308. III$_1$. **Felhoen** patented an explosive consisting of ten parts of nitro naphthaline with the crude ingredients of gunpowder, thus :—

Nitro-naphthaline - - - -	10 parts.
Saltpetre - - - -	75 „
Charcoal - - - - -	12½ „
Sulphur - - - -	12½ „

He stated that he obtained a mono-nitro-napthaline ($C_{10}H_7 . NO_2$) containing a small proportion of dinitro-naphthaline ($C_{10}H_6 . 2NO_2$) by digesting one part of naphthaline, with or without heat, in four parts of nitric acid (S.G. 1·40) for five days. (Spec. No. 2,266, 9.6.79.)

309. IV$_2$. **Fenton's Powder** consists of chlorate of potash, sugar, and yellow prussiate of potash in the proportions of 4 to 8 ozs. of each of the latter ingredients to 16 ozs. of damp chlorate. The mixture is made in the form of a stiff dough, being damped with lime water, gum water, or water. It is dried in an oven, cut up with brass knives, and sifted into various sizes of grains. The powder can be coloured with various colouring matters, which it is apparently proposed also to use in lieu of sugar. It is claimed as suitable for all kinds of small arms and ordnance and for blasting purposes. (Spec. No. 4,148, 17.12.73.)

310. III$_1$. **Filite.** The name given by the Italian Government to their smokeless powder. The composition is probably the same as Ballistite, whilst the name would tend to show that the powder is manufactured in the Cordite form.

311. III$_1$. **Fitch,** also **Reunert,** proposed a mixture consisting of 10 parts of nitro-glycerine to 90 parts of an absorbent composed of 73 parts of nitrate of soda, 12 parts of charcoal, 10 parts of sulphur, and 5 parts of starch. It is a sort of nitrate mixture with a little nitro-glycerine added. (Spec. No. 7,497, 22.5.88.)

312. **Fitch.** *See* STARCH POWDER.

313. **Flameless Securite.** *See* SECURITE.

314. **Flamboyure.** *See* PULVERIN.

315. **Fléron (de).** *See* PERTUISET.

316. **Fluorine.** *See* TURPIN (981).

317. III$_1$. **Fonite.** A dynamite with active base.

318. IV$_2$. **Fontaine's Powders** consist of picrate and chlorate of potash, and are intended for use in torpedoes and shells, but they are obviously very dangerous to manipulate, and caused a terrible explosion in Paris in 1869. (D., p. 739.)

319. **A.** III$_1$. **Forcite** has been described as a mixture of nitro-glycerine with cellulose, the latter being gelatinised by heating in water under considerable pressure. The American patent terms it a combination of nitro-glycerine with "an inexplosive gelatinising material and an "oxidating salt." Samples, however, examined in England consisted of a thin blasting gelatine with which were incorporated nitrate of potassium, wood meal, or pulp, and a little dextrine. Other samples contained nitro-glycerine, nitro-cotton, nitrate of sodium, and charcoal. A variety manufactured in America consists of a thin blasting gelatine incorporated with a mixture consisting of three parts of sulphur to 20 parts of wood tar and 77 parts of nitrate of sodium. To this mixture one per cent. of wood pulp is added to counteract the sticky qualities of the tar, while the latter is said to prevent the thin blasting gelatine from soaking into the base, which is thus used as a carrier rather than as an absorbent.

It is the invention of Captain J. M. Lewin, of the Swedish Army.

Forcite is defined as consisting of thoroughly purified nitro-glycerine, thickened by being combined with nitro-cotton and mixed or incorporated with wood meal and nitrate of potassium, in such proportions that the whole shall be of such character and consistency as not to be liable to liquefaction or exudation. It is therefore simply a gelatine dynamite. (Spec. No. 4,943, 27.11.80.)

Forcite—*continued.*

As made in Belgium, it has the following proportions:—

—	Extra.	Superieur.	No. 1.	No. 2.
	Per cent.	Per cent.	Per cent.	Per cent.
Nitro-glycerine	64·0	64·0	49·0	36·0
Nitro-cellulose	3·5	3·0	2·0	3·0
Nitrate of soda	—	24·0	36·0	35·0
Nitrate of ammonia	25·0	—	—	—
Wood meal	6·5	8·0	13·0	11·0
Magnesia	1·0	1·0	—	1·0
Rye-bran	—	—	—	14·0

320. III $_1$. **Forcite Gelatine,** a blasting gelatine containing 96 per cent. of nitro-glycerine. (Salvati.)

321. **Förster (Von).** *See* WOLFF and VON FÖRSTER.

322. **A.** II. **Fortis Explosive,** proposed by Heusschen, consists of:—

—	1.	2.	3.	4.	5.	6.	7.	8.	9.
	Parts.	Parts.	Parts.	Parts.	Parts.	Parts.	Parts.	Parts.	Parts.
Nitrate of potash	65·01	66·02	67·03	68·04	69·05	70·06	72·08	74·10	93·19
Sulphur	12·00	12·00	12·00	12·00	12·00	12·00	12·00	12·00	12·00
Lampblack	3·00	3·00	3·00	3·00	3·00	3·00	3·00	3·00	3·00
Tan	20·00	20·00	20·00	20·00	20·00	20·00	20·00	20·00	25·00
Sulphate of iron	1·39	2·78	4·17	5·56	6·95	8·34	11·12	13·90	27·80
Glycerine	0·92	0·92	0·92	0·92	0·92	0·92	0·92	0·92	1·84

The first five materials are powdered, wetted, and evaporated nearly to dryness. It is then completely dried, cooled, and the glycerine added to it. It is a black powder. It is claimed that various forms of nitro-glycerine are produced in the act of explosion in a blast-hole. (Spec. No. 1,024, 8.1.84.) The mixture, and the advantages claimed, are much the same as in the case of *Courteille's Powder, q.v.*

This Explosive has been licensed for importation only in the form of compressed cartridges thoroughly water-proofed.

323. A. II. **Fortis No. 2** consists of the same materials as fortis, with the addition of sulphuric acid and naphthaline. Analysis of a sample gave the following proportions:—

Nitrate of soda - - -	73·5 per cent.
Lampblack - - - - - -	3·0 ,,
Tan - - - - -	7·5 ,,
Sulphur - - - - - - -	8·5 ,,
Sulphate of iron - - -	1·0 ,,
Glycerine - - - - - -	3·0 ,,
Sulphuric acid - - -	1·5 ,,
Naphthaline - - - - -	2·0 ,,

In the manufacture of Fortis No. 2 most of the ingredients are mixed with water, and heated for some time, after which the two remaining ingredients, viz., glycerine and naphthaline, are added. As the result of this heating in presence of water the free sulphuric acid acts on the nitrate, liberating HNO_3, which in turn acts on the sulphate of iron and oxidises it to oxide of iron, being entirely decomposed and driven off if the heating is continued long enough. In this case no free acid is left, nor is any liable to be generated.

The explosive is authorised for importation only when it is free from acid and nitro-compound, and when it is thoroughly waterproofed.

324. A. III $_2$. **Fortisine** (originally termed Fortis Nos. 3 and 4) consists of a mixture of saltpetre, sulphur, and charcoal, with the addition of dinitro-benzol and resin (or dextrine). The amount of dinitro-benzol is not to exceed 4 per cent. of the finished explosive, and all the ingredients are to be thoroughly purified.

Analysis of a sample gave:—

Dinitro-benzol - - -	4·10 per cent.
Sulphur - - - - - -	6·30 ,,
Nitrate of potash - - -	74·80 ,,
Charcoal - - - - - -	14·40 ,,
Moisture - - - -	0·36 ,,

325. **Fournier** proposed as an explosive a mixture consisting of:—

Carbonate of lime - - -	125 parts.
Chlorate of soda - - - -	65 ,,
Urine, enough to cover the above when in a vessel.	

The mixture was evaporated nearly to dryness, and subsequently mixed with 35 parts of charcoal. This

Fournier—*continued.*
curious compound was to be used as a substitute for gunpowder. (Spec. No. 507, 21.2.70.)

326. III_1. **Fowler's Explosive** consists of a mixture of 20 parts of nitro-glycerine, 5 parts of charcoal, and 75 parts of an absorbent composed of 75 parts of nitrate of ammonia and 25 parts of anhydrous sulphate of soda. The last ingredient is prepared from sulphate of ammonia and nitrate of soda by a double decomposition. The object is to obtain a cheap combination of nitrate of ammonia in which the latter does not deliquesce. (T., p. 105.)

327. IV_2. **Fraenckel** places a cartridge of chlorate of potash in a waterproof cover inside a waterproofed nitrate of ammonium cartridge. The mechanical idea is similar to that of Favier, *q.v.* He also coats a nitrate or nitrates with two parts naphthaline to one part paraffin, stirring well together, and when required for use mixes with chlorate of potash. (Spec. No. 13,789 27.7.89.)

328. III_2. **Franke** has patented a process consisting of the addition of an agglutinating material, such as collodion, drying oil, gum, or solution of silicate of potash to a mixture of the nitro-derivatives of benzol, phenol, cresylol, naphthaline, and naphtol with solid oxidising agents. The explosive so formed is compressed into cartridges. (Fr. Spec. No. 179,452, 4.11.86.)

329. **Freeden.** *See* VON FREEDEN.

330. II. **Freiberg Powder** consists of:—

Nitrate of soda	61·66 parts.
Sulphur	17·25 ,,
Charcoal	17·35 ,,

(D., p. 609.)

331. IV_2. **French Green Powder** consists of:—

Chlorate of potash	14 parts.
Picric acid	4 ,,
Yellow prussiate of potash	3 ,,

Each ingredient is separately dried, and finely powdered, and the mixture is made in glass vessels, or, preferably, in wooden vats with wooden balls.

332. **Frost.** *See* WETTER.

333. V_1. **Fuch's Powder** consists of a mixture of chlorate of potash, finely ground tortoise or turtle shell, saltpetre, sulphur, and charcoal. (T., p. 101.)

334. III_1. **Fulgurite** consists of nitro-glycerine mixed with some coarsely ground farinaceous substance, preferably corn meal, in varying proportions. (T. p. 102.)

335.V_1. **Fulminates,** *see* p. 44.

336. V_1. **Fulminate of Copper** forms highly explosive green crystals, and is obtained by boiling fulminate of mercury (or fulminate of silver, in which case a greenish blue powder is obtained) with copper and water.

There are also double fulminates of copper and ammonium, and of copper and potassium.

337. **A.** V_1. **Fulminate of Mercury** is the fulminate by far the most in practical use. It is produced by the action of alcohol on mercury dissolved in nitric acid. When pure it is white, but is often greyish or yellowish. It is extremely explosive, and has very violent local action. It explodes with slight friction or percussion, or when heated to about 360° F. When thoroughly wet it is inexplosive. The larger the crystals the more sensitive is the fulminate.

338. **A.*** V_1. **Fulminate of Silver** is used in bon-bon crackers and other toy fireworks in minute quantities. It is prepared by much the same process as fulminate of mercury, *i.e.*, by the action of alcohol on the nitrate of the metal. It is more sensitive than fulminate of mercury, and can, it is stated, even be exploded under water by friction with a glass rod. It crystallizes in small white needles. It forms an intensely sensitive and powerful fulminate with ammonium, and several other.

---

* Licensed for toy fireworks only.

Fulminate of Silver—*continued.*
metals, that with hydrogen is the acid fulminate of silver.

339. V_1. **Fulminate of Zinc** is obtained by boiling fulminate of mercury with zinc, and it forms an acid fulminate with hydrogen, and double fulminates with many metals.

340. III_1. **Fulminatine** is a mixture of as much as 85 per cent. of nitro-glycerine with a substance "prepared by a chemical process." Burnt in the open it leaves a slight residue rich in carbon. (D., p. 728.)

341. V_2. **Fulminating Gold** is a violently explosive buff precipitate formed when ammonia is added to ter-chloride of gold. It is an ammonium compound, in which part of the hydrogen of the ammonium is replaced by gold.

342. V_2. **Fulminating Platinum** is a violently explosive black precipitate formed by mixing ammonia with a solution of binoxide of platinum in dilute sulphuric acid. It is, generally speaking, of the same chemical nature as fulminating gold.

343. V_2. **Fulminating Silver** (as distinguished from Fulminate of Silver) is a similar compound to the last, obtained by the action of ammonia on oxide of silver. It is very violently explosive.

344. III_1. **Fulmine Dynamite** is similar to Jupiter, Neptune, Titan, Vulcan, and other powders. (P. & S. 364.)

345. V_1. **Fulminurates** are feeble explosives obtained by the action of chlorides on fulminates. Fulminuric acid has the formula $C_3N_3H_3O_3$ in which one atom of hydrogen may be replaced with one atom of a metal to form the fulminurate.

346. **A.** VI_3. **Fuzes for Shells** are licensed only when they are of such strength and construction that the explosion of one fuze will not produce an explosion en masse of other like fuzes.

347. II. **Gacon's Powder** consists of 69 parts of potassium, or sodium nitrate, and 19 parts sulphur. The ashes of dead leaves are added, and finally a solution of tannin in water. It is said to require a temperature of 900° F. to ignite it, and to be absolutely insensible to percussion. (Spec. No. 5,735, 13.12.83.)

348. III_1. **Gacon** has also patented a dynamite with a base of "ramie"* in a fine state of division, and the mixture of this dynamite with the powder (347) above described. (French Spec. 158,724, 19.3.86.)

349. III_2. **Gaens** mixes dinitro-cellulose gelatinised by acetic ether with nitre and bumate of ammonia. The bumate is prepared from peat washed and boiled in solution of carbonate of soda. (German Patent, D.R., 48,933, 19.3.89.)

350. **Gaens.** *See* AMIDE POWDER.

351. II. **Gallaher's Powder** consists of:—

Nitrate of soda, or ,, potash	70	to 80	parts.
Sulphur	6	to 12	,,
Sulphate of iron	1	to 3	,,
Charcoal	8	to 16	,,
Sulphate of copper	0·5	to 1	,,
Ground bark	8	to 14	,,

It is claimed to be safer than ordinary blasting powder, and not to be liable to ignition by friction or percussion. (T., p. 106.)

352. **Gallic Powder.** *See* HORSLEY.

353. **Garside.** *See* HARRISON.

354. **A.** III_2. **Gathurst Powder** is defined as consisting of a mixture of any of the following substances:—Nitrates of potash, soda, and ammonia, sulphates of magnesia and ammonia, chloride of ammonia, colouring matter such as lampblack or charcoal, nitro- and chloro-nitro compounds of benzol, toluol, and naphthaline. There is a proviso that

* This is China-grass or rhea, a fibrous species of nettle.

Gathurst Powder—*continued.*

the amount of chlorine in the finished explosive shall not exceed 2 per cent.

Analysis of a sample gave :—

Dinitro-benzol - - -	16·52 per cent.
Nitrate of ammonia - -	83·40 „
Moisture - - - -	0·08 „

355. II. **Gaüs Powder,** a mixture of saltpetre, charcoal, and nitrate of ammonia. (P. & S. 377.)

356. IV_2. **Gas Powder,** a simple modification of Melland's Paper Powder. (D. 614.)

357. **A.** III_1. **Gelatine Dynamite** occupies a place mid-way between blasting gelatine and dynamite. It consists of a thin blasting gelatine mixed with other substances. Two varieties are licensed: No. 1 contains cotton, charcoal, or "such other ingredients as may for the time being be "sanctioned by a Secretary of State;" No. 2 consists of No. 1 mixed or incorporated with nitrate of potash or other nitrates. The varieties practically in use contain nitro-glycerine, nitro-cotton, nitrate of potash, and wood meal. They resemble blasting gelatine very closely in appearance, and it requires practice to distinguish them apart. Both come under the No. 2 class and differ in grade only. One contains about 80, the other 60 per cent. of explosive. The last-named is known as gelignite.

358. III_2. **Gelbite** is a species of nitro-paper with the addition of picric acid. *See* EMMENS. (P. & S. 386.)

359. **A. Gelignite.** *See* GELATINE DYNAMITE.

360. II. **Gemperle** has patented an explosive composed as follows :—

Saltpetre - - - -	73 per cent.
Bran - - - - - -	8 „
Charcoal - - - -	8 „
Sulphur - - - - -	10 „
Sulphate of magnesia - -	1 „

The mixing is done with the wet ingredients. This explosive is made in Switzerland and is termed Amidogène. (O. G. Spec. No. 2,407, 22.5.82.)

361. III_2. **Germain** has patented the use in the manufacture of explosives of cellulose extracted from cocoanut fibre. (Fr. Spec. 192,181, 31.7.88.)

362. **A.** VI_2. **German Spills** are paper tubes containing a little fine-grained gunpowder, and are primed at one end with touch paper. They are used as fuzes instead of "straws" for firing charges of gunpowder in blast-holes.

363. III_1. **Giant Powder** is a term loosely used as a synonym for dynamite, but the explosive made in California under this name is essentially a lignin-dynamite containing nitrate of sodium. *See also* EXTRA.

364. III_1. **Giant Powder No. 2** consists of:—

Nitro-glycerine	40 per cent.
Nitrate of soda „ potash	40 „
Resin	6 „
Sulphur	6 „
Infusorial earth	8 „

Other nitrates may be substituted for those above named, and other carbonaceous substances for the resin. It is practically Dynamite No. 2.

There are in all seven grades of giant powder containing from 75 to 20 per cent. of nitro-glycerine.

365. **Gigantic.** *See* HELLHOFFITE.

366. **Gilles.** *See* NITRO-MOLASSES.

367. **Girard** (Aimé). *See* HYDRO-NITRO-CELLULOSE.

368. III_1. **Girard** employed sugar as an absorbent for nitro-glycerine. Millot and Vogt proposed the same mixture.

369. **Glaser.** *See* DUTTENHOFER.

370. III_1. **Glonoine,** an early name or nitro-glycerine.

371. III$_1$. **Glukodine** is a whitish liquid produced by the nitration of a saturated solution of cane sugar in glycerine. Free sugar dissolves in it, and it is soluble in ether. Two sorts of glukodine powder are made, white and black. Samples showed the following composition:—

—	White.	Black.
	Per Cent.	Per Cent.
Matter soluble in ether (glukodine) - -	36·40	34·24
Free sugar - - - - -	8·40	8·76
Soda salts (mostly nitrate) - -	31·20	37·84
Nitro-cellulose - - -	23·36	—
„ „ and charcoal - - -	—	19·31

Another analysis of the same samples showed that the glukodine in the white powder consisted of 33·19 parts of nitro-glycerine to 3·21 parts of nitro-saccharose, and in the black powder of 30·23 and 4·03 parts of the same ingredients respectively. Subsequent experiments appeared to show that glukodine was simply a mechanical mixture of those ingredients, as the nitro-glycerine could be readily evaporated off from the nitro-saccharose. (M. No. II., p. 445 *et seq.*)

372. **Glycero-nitre.** *See* FORTIS.

373. III$_2$. **Glycero-nitre.** Under this name also, a French patent claims a mixture of an alkaline nitrate, sulphur, charcoal, wood meal, and a nitrated hydrocarbon such as dinitro-benzol. (Fr. Spec. No. 214,741, 8.7.91.)

374. **Glycero-pyroxiline.** *See* CLARK.

375. **A.** III$_1$. **Glyoxiline** consists of nitrated gun-cotton impregnated with nitro-glycerine, and is due to Sir F. Abel, Bart. A similar material was submitted by Mr. Broderson for license in 1885 and approved, but no steps have since been taken for its introduction. (Spec. No. 3,652, 24.12.67, and D. 725.)

376. V_1. **Goetz' Powder** is a mixture of an explosive base with glucose (grape sugar $C_6H\ O_6$), uncrystallisable sugar, or syrupy solution to prevent premature or accidental discharges. The proportions recommended are:—

Chlorate of potash - - -	10 parts.
Solution of glucose - - -	10 ,,
Charcoal in powder - -	3 ,,
Sulphur - - - -	2 ,,
Amorphous phosphorus - -	1 ,,
Picrate of lead - - -	3 ,,

It is claimed that the compound will stand all ordinary shocks, handling, and transport, without danger of explosion. It burns with a bright light when unconfined. (T., p. 105.)

377. IV_2. **Gomez' Powders** are mixtures of chlorate and prussiate of potash, with sugar (acetate) of lead, and nitrate of lead, or nitrate of iron. (T., p. 101.)

378. **Gomme.** *See* GELATINE DYNAMITE.

379. II. **Goodyear** proposed to employ india-rubber or gutta-percha (apparently in lieu of charcoal) in the manufacture of gunpowder. (Spec. No. 1,703, 26.7.55.)

380. IV_1. **Gotham's Explosive** consists of:—

Chlorate of potash -	10 parts.	} 70 to 65 per cent.
Nitrate ,, ,, -	2 ,,	
Powdered oak bark -	5 ,,	
Nitro-glycerine - - -		30 to 35 ,,

(T., p. 105.)

381. **Gottheil.** *See* RHENISH DYNAMITE CO.

382. IV_2. **Graham's Powder** consists of:—

Yellow prussiate of potash - -	16 parts.
Chlorate of potash - -	40 ,,
White sugar - - - -	20 ,,
Red lead - - - -	1·25 ,,

It is claimed that the red lead obviates the danger of explosion, and that the compound can be used when wet and doughy. (T., p. 106.)

383. III. **Grakrult** (Skoglund) is a smokeless powder with which experiments have been made by the Swedish navy.

384. **Granatina.** *See* SALA.

385. **A.** III $_2$. **Granulite** is defined as consisting of nitro-cotton mixed with a nitrate or nitrates, with or without the addition of charcoal and paraffin. It is a yellowish waxy-looking material in large irregular grains, coated with paraffin. Analysis of a sample gave:—

Nitro-cotton	46·28	per cent.
Nitrate of barium	45·48	,,
Paraffin	7·84	,,
Moisture	·40	,,

386. III $_1$. **Graydon** soaks woollen or cotton cloth in nitro-glycerine till saturated. The whole is coated with paraffined paper cemented to the cloth, or with shellac, or any coating which will prevent the exudation of the nitro-glycerine. The cloth may be in sheets or ribbons, and charges are made by rolling up the fabric into cylinders. When in ribbons the charges may be made up by placing the tape-like discs on each other. (M. XXI., p. 491, *q.v.*, for other devices by the same inventor.)

387. III $_1$. **Green** absorbs nitro-glycerine in a mixture of keiselguhr and charcoal or carbonised starch, and claims that the dynamite so made will not exude in water.

388. **Green Powder.** *See* FRENCH GREEN POWDER.

389. I. **Greene** reproduced Drayson's process, with the addition of mixing the ingredients in a vacuum or partial vacuum. (T., p. 103.)

390. **A.** III $_1$. **Greener's Powder** consists of a mixture of nitro-cellulose with nitro-benzol, with the addition of colouring matter, consisting of graphite, lampblack, or other suitable material.

391. **Gregory.** *See* WARD and GREGORY.

392. **Grenadine.** *See* GACON (347) and SALA.

393. **Grey Powder.** *See* GRAKRULT.

394. **Griess.** *See* CHROMATE DE BENZINE.

395. **Grises de Paulilles.** *See* DYNAMITE (254).

396. **Grisoutine Comprimée.** *See* BENDER.

397. **Grisoutine,** as made in France, consists of thin blasting gelatine with various proportions of nitrate of ammonia added. *See* WETTER.

398. III $_1$. **Grisoutite** (*grisou*, fire-damp) is another name for Wetter Dynamite. As proposed by Faille it consists of :—

Nitro-glycerine - - -	42 to 45 per cent.
Kieselguhr - - - -	11 to 12 ,,
Sulphate of magnesia - -	47 to 43 ,,

(P. & S. 425.)

399. **Grouselle.** *See* PYROXILITE.

400. III $_1$. **Grüne** has proposed to render dynamite unaffected by moisture by mixing animal or vegetable charcoal with the kieselguhr, or by charring the latter mixed with organic substances. (Spec. No. 16,116, 23.11.87. Fr. Spec. No. 187,345, 1.12.87.)

401. **Grüson.** *See* HELLHOFFITE.

402. III $_1$. **Guinard** manufactures a smokeless powder in the form of greyish, semi-transparent tablets of irregular shape. It is probably a gun-cotton powder.

403. **A.** III $_2$. **Gun-cotton.** For full particulars of this explosive, *see* NITRO-CELLULOSE, p. xxxiii. For licensing purposes in this country gun-cotton has been distinguished from collodion cotton by the following definition :—

Gun-cotton, consisting of thoroughly purified nitro-cotton (*a*), of which not more than 15 per cent. is soluble in ether alcohol, or (*b*) which contains more than 12·3 per cent. of nitrogen ; whether with or without carbonate of calcium.

404. **A.** VI $_2$. **Gun-cotton Fuzes** are licensed as consisting of cases of metal or other suitable material, containing not more than two drams of gun-cotton.

405. **A. I. Gunpowder.** *See* INTRODUCTION, p. xx.

406. **A. VI$_2$. Gunpowder Fuzes** are licensed as consisting of cases of metal or other suitable material, containing not more than two drams of gunpowder.

407. **II. Güttler** proposed to make cartridges of compressed blasting powder by cementing the grains together with dextrine. He used a brown-red charcoal made from wood, free from resin, at a temperature of about 290° C. This charcoal was said to have the formula $C_8H_4O_2$. (M., No. V., p. 749.)

408. **V$_1$. Hafenegger's Powders** consist of the following compounds:—

—	No. 1.	No. 2.	No. 3.	No. 4.	No. 5.	No. 6.	No. 7.
	Parts.	Parts.	Parts.	Parts.	Parts.	Parts.	Parts.
Chlorate of potash - -	9 or 6	2	4	4	1	11	1
Sulphur - - -	¼ ,, 1	—	1 (or sugar)	¼	—	¾	—
Charcoal - - - -	¼ ,, 4	—	¼	½	—	¾	1
Yellow prussiate of potash	—	1	1	—	—	—	—
Sugar - - -	—	1	1 (or sulphur)	4	1	—	—

He recommends for use with the above a self-igniting liquid consisting of solution of one or two parts of phosphorus in two of bisulphide of carbon, the time before explosion takes place, being regulated by the amount of saturation of the powder with the liquid, the average period being about half-an-hour. He also proposes simply a mixture of chlorate of potash with the liquid. (Spec. No. 2,865, 17.9.68.)

The danger and uncertainty of the propositions are obvious.

409. **V$_1$. Hahn** proposed a priming composition for needle guns consisting of:—

Chlorate of potash - - -	20 parts.
Picric acid - - - - - - -	5 ,,
Amorphous phosphorus - - -	5 ,,
Tersulphide of antimony - - - -	1 ,,

made into a paste with water and gum. (Spec. No. 961, 30.3.67.)

410. V_1. **Hahn's Powder** consists of:—

Chlorate of potash - - -	367·5 parts.
Tersulphide of antimony - - -	168·3 ,,
Spermaceti - - - -	46·0 ,,
Charcoal - - - - - - -	18·0 ,,

The idea is to add the chlorate of potash only when required for use, the mixing being done by sieves. The addition of the spermaceti is claimed to give safety against explosion by friction. (Spec. No. 960, 30.3.67.)

411. V_1. **Hall's Powder** consists of:—

Chlorate of potash - - -	47 parts.	
Ferrocyanide of potassium - -	38 ,,	
Sulphur (or other chemical equivalent)	5 ,,	(about).

The ingredients are pulverised and mixed in water, or "water and nitric acid." To the evaporated mixture are added 10 parts of caoutchouc, sometimes slightly impregnated with bisulphide of carbon. The whole is mixed, pressed, and granulated. (Spec. No. 1,062, 28.4.63.)

412. **Haloxyline.** *See* BLECKMANN.

413. II. **Haloxyline,** proposed by Anders and Fehleisen, consists of:—

Saltpetre - - - - -	75 per cent.
Sawdust - - - - - - - -	15 ,,
Charcoal - - - - -	$8\frac{1}{4}$,,
Red prussiate of potash - - - -	$1\frac{3}{4}$,,

It is made in Austria. A variety made in Hungary contains nitrate of soda. (D. 602.)

414. IV_2. **Hannan** proposed a mixture of yellow or red prussiate, nitrate, and chlorate of potash, with vegetable or animal charcoal. Paraffin or other oleaginous or fatty substance or gums are introduced as a binding material. (Spec. No. 4,846, 12.10.82, and No. 5,323, 8.11.82.)

Subsequently he added ferric oxide to sharpen the explosion and render it local. (Spec. No. 5,986, 15.12.82.)

415. **Hardingham** has patented the employment in blasting cartridges of a non-inflammable liquid (such as liquified ammonia or carbonic acid), or of an inflammable liquid (such as alcohol, ether, or benzol) in combination with gunpowder, dynamite, or other explosive. (Fr. Spec. No. 164,996, 25.10.84.)

416. II. **Hardy's Powder** consists of:—

	"000."	"00."	"0."
	Parts.	Parts.	Parts.
Nitrate of soda - - - -	46	78·58	76·66
„ potash - - - - -	21	18·20	19·83
Sulphur - - - - -	15	10· 5	13·00
Charcoal - - - - -	15	15· 0	17·00
Sugar and tartar - - - - -	2·65	6· 3	2·80
Moisture - - - - -	0·35	1·60	0·71

(O.G.)

417. V_1. **Harrison's Powders** consist of mixtures of carbon, sulphur, clubmoss, and chlorate of potash. The following proportions are given:—

Chlorate of potash -	7 parts, or	for	rifle shooting	11	parts.
Starch - -	1 „	„	„	1	„
Charcoal - - -	1 „	„	„	1	„
Sulphur - -	1 „	„	„	2	„
Clubmoss (lycopodium clavatum) - -	— „	„	„	¼	„
Coal - - -	— „	„	„	1	„
Soot from coal -	— „	„	„	½	„

(Spec. No. 2,642, 29.10.60.)

Subsequent specifications give:—

Chlorate of potash -	56 parts	12 parts	(or chlorate of soda).
Ferrocyanide of potash	28 „	4 „	
Starch - - - -	4 „	2 „	
Sulphur - -	7 „	— „	
Charcoal - -	5 „	1 „	(or $2\frac{1}{2}$ parts of coal or cannel).
Nitrate of potash -	— „	6 „	

(Spec. Nos. 2,233, 6.9.61, and 305, 5.2.62.)

418. II. **Harrison** has also proposed the following:—

Saccharum (saccharine matter) - -	50 per cent.
Saltpetre - - - - - -	48 „
Clubmoss (lycopodium clavatum) - -	2 „

(Spec. No. 1,786, 24.7.60.)

419. **A.** III_2. **Hauff** proposes to use trinitro-resorcin $C_6H(NO_2)_3(OH)_2$ as an explosive, the crystals being granulated in a suitable manner. This material differs from picric acid in containing one more atom of oxygen. It is stated, however, to be a less violent explosive. (Spec. 9,798, 19.5.94.)

419. IV_2. **Hart** proposed chlorate of potash granulated and impregnated with a saccharine solution, or other suitable hydro-carbon liquid. (Spec. No. 9,164, 23.6.88.)

420. **Hay, Merricks, & Co.** *See* ROSSLYN POWDER.

421. **Heath.** *See* WETTER.

422. II. **Hebler Powder.** A so-called smokeless powder, which was manufactered in Switzerland. Its composition was found by analysis to be :—

Nitre - - - -	64·41 per cent.
Ammonium nitrate - - -	16·35 ,,
Sulphur - - - -	9·80 ,,
Charcoal - - - - - -	11·84 ,,
Moisture - - - - -	·92 ,,

The powder does not appear to have a greater tendency to absorb moisture than black powder.

423. III_1. **Hecla Powder,** an American form of lignin-dynamite made in seven grades, containing from 75 to 20 per cent. of nitro-glycerine, with the addition of nitrate of sodium.

424. **Heick.** *See* THUNDER POWDER.

425. **Heliophanite.** *See* PANCLASTITE.

426. III_1. **Hellhoffite** is a mixture of nitro-petroleum or nitro-tar with nitric acid. One form of it proposed by Gruson has been tried in shells, and consists of meta-dinitro-benzol and nitric acid. The two substances are placed in separate receptacles in the shell, and are automatically mixed during the flight, or on the impact of the shell, as the case may be. (Spec. No. 1,315, 23.10.79; Nos. 1,285–7, 27.3.80; and No. 2,775, 7.7.80.)

The proportions of Hellhoffite for blasting are given as :—

Dinitro-benzol -	1 part to 1½ parts of nitric acid.
Or nitro-benzol -	1 ,, 2½ ,, ,,

427. R. III_2. **Hengst** has patented under the name of Hengstite :—(1.) The employment of permanganate of potash for oxidising nitrated straw pulp which is afterwards treated with a bath of carbonate of potash. (2.) A gelatinous substance obtained by boiling linseed with the addition of a small quantity of dextrine for reducing the

Hengst—*continued.*

substance to paste. (3.) A solution prepared by dissolving in sulphuric ether the powder or its siftings. This solution serves for giving the grains an impermeable coating.

The powder is a fine fibrous substance which is granulated for military and sporting purposes, and compressed for blasting purposes. (Spec., No. 13,656, 21.9.88. Fr Spec. No. 194,803, 17.12.88.)

428. III $_2$. **Heraklin** consists of picric acid, nitrates of potash and soda, sulphur, and sawdust (of oak or other hard wood). It is manufactured as follows:—In 36 parts of boiling water half a part each of picric acid and saltpetre are dissolved, and 15 parts of sawdust are impregnated with this solution. The impregnated sawdust is dried, and to every 10 parts thereof are added 17·5 parts each of nitrates of potash and soda, and 7·5 parts of sulphur. All the ingredients are finely powdered. Compressed sticks or bars of the same material are made by treating the material throughout in a damp state. (Spec. No. 4,155, 1.12.75.)

429. IV $_1$. **Hercules Powder,** an American form of dynamite containing from 75 to 20 per cent. of nitro-glycerine. It is made in 10 grades of which two are constituted as follows:—

—	No. 1.	No. 2.
	Per Cent.	Per Cent.
Carbonate of magnesia - - - -	20·85	10·00
Nitrate of potash - - - - -	2·10	31·00
Chlorate - - - - - -	1·05	3·34
White sugar - - - - -	1·00	15·66
Nitro-glycerine - - - - -	75·00	40·00

(T., p. 103.)

Some varieties consist of wood pulp, nitrate of soda, carbonate of magnesia, and nitro-glycerine. Another variety is called No. 1 Extra Hercules Powder, and is identical with Nitro-Magnite, *q.v.*, containing 80 per cent. of nitro-glycerine to 20 per cent. of magnesia alba.

430. II. **Herculite,** a yellowish grey powder composed of saw-dust, camphor, saltpetre, and other substances. It is used for blasting.

431. **Heusschen.** *See* FORTIS and GLYCERONITRE.

432. III_1. **Heusschen** has also patented under the name of benzo-glycero-nitre the production of nitro-glycerine and nitro-benzol by the combination of sulphuric acid, a sulphate, nitrate of soda or potash, glycerine, and coal-tar oil. (Fr. Spec. 182,050, 8.3.87.)

433. III_1. **Hill's Powder** consists of a mixture of nitro-glycerine and a silicious powder prepared by a precipitation from a solution of silicates.

434. IV_1. **Himley** mixes :—

Chlorate of potash - - -	45 per cent.
Nitrate of potash - - -	35 ,,
Coal tar - - - - -	20 ,,

The tar is dissolved in petroleum ether, and the latter evaporated off. It is made in Russia. (O. G.)

435. II. **Himly** proposed a gunpowder in which hydrocarbons precipitated from solutions in naphtha take the place of the charcoal and sulphur of ordinary powder. The powder so made is said to be quite waterproof. (M., IV., 298.)

436. III_1. **Hinde** proposed an explosive consisting of :—

Nitro-glycerine - - - -	64 per cent.
Ammonium citrate ($C_6H_7(NH_4)O_7$) -	12 ,,
Ethyl palmitate ($C_{18}H_{36}O_2$) - - -	0·25 ,,
Calcium carbonate - - -	0·25 ,,
Sodium carbonate - - - -	0·50 ,,
Coal - - - - -	23 ,,

(M., No. V., p. 731.)

437. V_1. **Hochstätter's Compound** consists of a mixture of chlorate of potash or of lead, nitrate of potash or of soda, and charcoal or sulphur, or a metallic sulphide. The mixture is dissolved in water, in which paper or vegetable matters are steeped so as to render them explosive. (Spec. No. 2,869, 17.12.69.)

438. I. **Hodge** injected steam into the mixed ingredients of gunpowder with a view of dissolving the saltpetre, and thoroughly combining the ingredients. (Spec. No. 2,227, 22.8.57.)

439. **Hood.** *See* MOSENTHAL.

440. II. **Hope** substitutes for the whole or a portion of the charcoal in gunpowder, starch, flour, sugar, or other like organic material containing carbon, oxygen, and hydrogen. Also bitumen or other solid hydro-carbon to ensure more perfect combustion. (Spec. No. 14,914, 12.11.84.)

441. III_2. **Hornite,** the name given to an experimental smokeless powder composed of nitro-cotton dissolved in acetone or acetic ether. The material was pressed into thin sheets and cut into tablets or ground into powder. Other ingredients were sometimes added.

442. IV_2. **Horsley's Powder** consists of a mixture of finely powdered chlorate of potash and gall nuts in the proportions of three to one. The powder is granulated by passing it through a sieve in a damp state. It is stated to explode at 430° F., and its force to be "five times that of gunpowder." (Spec. No. 1,193, 19.4.69, and No. 1,885, 22.6.72.)

443. IV_1. **Horsley** has also proposed :—

(1) The admixture of three parts of nitro-glycerine with eight parts of alum or sulphate of magnesia, finely powdered and sifted.

(2) The addition of 20 per cent. of nitro-glycerine to his chlorate of potash and nut gall powder. *See* (442).

Two varieties were licensed about 1872, and specified as containing nitro-glycerine thoroughly mixed with not less than 75 per. cent. of one of the following powders, pulverised or granulated :—

—	A.	B.
Chlorate of potash	3	6
Nut galls	1	1
Wood charcoal	...	1

444. V_1. **Howard's Powder,** a name given to fulminate of mercury, discovered by Howard in 1800. (Salvati.)

445. **R.** V_1. **Howittite** is a mixture of picric acid with chlorate of potash and nitrate of soda. It is very sensitive, and is unstable. The action of the picric acid on the chlorate at 100° F. is very powerful.

446. III_1. **Hudson's Explosive.** A sort of blasting gelatine employed in America (1889) as a bursting charge for shells. It consists of an intimate mixture of nitro-glycerine with nitro-cellulose, the latter being previously dissolved in acetone or other solvent. (P. & S. 468.)

447. III_2. **Huetter's** blasting compound consists of the same ingredients as Tonite.

448. I. **Hunt** proposed to mix the ingredients of black powder with sufficient water to form a thin paste, and to combine them in this condition in a revolving drum suitably fitted. (Spec. No. 3,775, 12.12.72.)

449. VI_2. **Hunter's Patent Miner's Fuzes** are much the same as German Spills, *q.v.*, but are waterproofed or varnished outside. One end of the tube is primed with saltpetre or sulphur, or both. The saltpetre only is best in fiery mines. The manufacture is not at present carried on.

450. III_2. **Hydronitro-cellulose,** prepared by Aimé Girard by nitrating cellulose, which has been previously disintegrated by means of hydrochloric or sulphuric acid. (P. & S. 474.) *See* CROSS, BEVAN, and BEADLE.

451. III_1. **Indurite** consists of gun-cotton freed from the lower nitro-celluloses by treatment with methyl-alcohol, and mixed with a liquid nitro-compound, such as nitrobenzol. The mass is incorporated in a mill or between rolls, and is then cut or moulded into suitable forms. The inventor (Munroe) prefers to use from 9 to 18 parts of nitro-benzol to 10 parts of gun-cotton. Suitable oxidising salts may be added. The material can be made of the consistency of bone or ivory, by treatment with hot water or steam. (Spec. No. 580, 11.1.93.)

452. A. VI$_2$. **Instantaneous Fuze.** Under this name is included any preparation of gunpowder, yarn, and protective coating which is not a safety fuze and which does not contain its own means of ignition.

453. V$_2$. **Iodide of Nitrogen** ($N.H.I_2$.), a violet powder, which, when dry, can be exploded by touching it with a feather. It is obtained by the action of ammonia on iodine.

454. III$_2$. **J. Powder,** a smokeless sporting powder, manufactured in France. It consists of:—

Gun-cotton - - - - -	83 per cent.
Bichromate of ammonia - -	17 ,,

(P. & S. 804.)

455. **Jahnite.** *See* JOHNITE.

456. **Jaline.** *See* JOHNITE.

456[a]. A. **Jaroljmek** proposes to ignite blasting charges in mines by means of heat developed by the action of the water used for tamping, on quick-lime. (Spec. No. 13,714, 16.7.94.)

457. **J. B. Powder.** *See* JOHNSON-BORLAND POWDER.

458. **J. C. P. Powder.** *See* PLASTOMENITE.

459. V$_1$. **Johnite** (also called Jahnite and Jaline) consists of:—

Nitrate of potash - - -	75 per cent.
Sulphur - - - - - -	10 ,,
Lignite - - - - -	10 ,,
Picrate of soda - - - -	3 ,,
Chlorate of potash - - -	2 ,,

("Mining Journal," 18.6.87.)

This explosive is made in Austria. The proportions sometimes vary from those given, and picric acid and calcined soda are added in small quantities.

460. V$_1$. **Johnson** proposed to employ amorphous phosphorus with metallic salts in the manufacture of fulminating powder. His proposal and proportions are identical with those of Alexander, *q.v.* (Spec. No. 2,377, 10.10.56.)

461. **A. III$_2$. Johnson-Borland Powders.** In these powders dinitro-cellulose or lower forms of nitro-cellulose are employed, impregnated with barium or potassium nitrates and incorporated with charcoal or other carbonaceous material. Two examples given in the specification consist of the following :—

—	For Military Arms.	For Sporting Arms.
	Per Cent.	Per Cent.
Nitro-cellulose - - -	50	50
Potassium nitrate - -	40	22
Barium nitrate - - -	—	25
Torrefied starch or lampblack	10	3

The powders are formed into grains or blocks, and impregnated with a solution of camphor and phenol, or camphor alone, in a suitable volatile solvent, in the proportion of one part of camphor (or camphor and phenol) in five parts of solvent to ten parts of the powder. The solvent is driven off at a gentle heat, and the camphor subsequently driven off at a temperature not exceeding 100° C.

It is claimed that by this method powders can be produced of any required degree of hardness and density, thereby regulating the energy of action of the explosive; and it is stated that "these results are obtained not by " the presence of camphor in large or small quantities in " the finished explosive, but by a remarkable gelatinising " (and perhaps some other) actions exerted by the camphor " upon the nitro-cellulose when these are heated together " at varying temperatures up to 100° C., whereby the " hardness and density of the explosive may be regulated " at will by the proportion of camphor used." The use of ordinary gun-cotton is expressly barred on account of the elements of uncertainty and danger introduced by it. (Spec. No. 8,951, 24.7.85.)

These powders have been licensed for manufacture by the E. C. Company, under the name of "E. C. Powder Company's (sporting or rifle) powder J. B. Patent."

462. **Johnson.** *See* REID and JOHNSON.

463. **Jolly.** *See* WIGFALL.

464. III_1. **Jones' Dynamite.** This explosive has been authorised in the Colony of Victoria. It consists of a low grade dynamite (35 per cent. nitro-glycerine), the absorbent being a mixture of kieselguhr and sulphate of lime.

465. III_1. **Judson Dynamite** is a mixture of nitro-glycerine, nitrate of sodium, and bituminous coal. The per-centage of nitro-glycerine is small.

466. III_1. **Judson Powder** is an American form of dynamite, made in four grades, containing from 5 to 20 per cent. of nitro-glycerine. It is a mixture of nitro-glycerine with various explosive absorbents, but differs from the usual form of such mixtures in that the grains of the absorbent are coated with some combustible substance offering resistance to the absorption of nitro-glycerine or water. The object is "to save and render effective that proportion of "nitro-glycerine which, with ordinary absorbents, is so "closely absorbed and taken up that it is rendered "practically inexplosive." Hence a comparatively very small proportion of nitro-glycerine will, it is stated, give a powerful explosive compound. The following is given as an example of such an absorbent:—

Sulphur - - - -	15 per cent.
Resin - - - - - -	3 ,,
Asphalt - - - -	2 ,,
Nitrate of soda - - -	70 ,,
Anthracite - - - -	10 ,,

The three first ingredients are melted together and well stirred, and the nitrate of soda and coal, dried and powdered, are stirred in till thoroughly mixed, the stirring to be continued gently till the grains cease to adhere to each other. The nitro-glycerine can be added to the finished mixture as desired. (T., p. 105.)

The composition of the R.R.P. grade, the one most commonly used, is:—

Sodium nitrate - - -	64 per cent.
Sulphur - - - - -	16 ,,
Cannel coal - - - -	15 ,,
Nitro-glycerine - - - -	5 ,,

Practically this is a sort of crude gunpowder, containing a little nitro-glycerine. It requires a primer of more powerful explosive to develop its full powers.

467. III_1. **Jupiter Powder,** a name given to a No. 2 Dynamite, similar to Vulcan or Neptune Powder.

468. IV_2. **Justice** proposed (1888) a mixture of nitrate and chlorate of potash, kneaded with paraffin or naphthaline.

469. **Kalk.** *See* ATLAS.

470. III_1. **Kadmite** consists of :—

Nitro-glycerine - - -	20 parts.
Nitrate of soda - - - -	56 ,,
Sulphur - - - -	10 ,,
Charcoal - - - - -	4 ,,
Ligneous matter - - -	4 ,,

(P. & S. 495.)

471. IV_2. **Keil's Explosive** consists of nitro-glucose (dextro-glucose made from starch) compounded with nitrate and chlorate of potash and prepared vegetable fibre. (T., p. 107.)

472. V_1. **Kellow** and **Short** proposed the following mixtures :—

Nitrate of soda - -	30 parts or	36 parts.
,, potash - -	8 ,,	4 ,,
Chorate of potash - -	12 ,,	6 ,,
Sulphur - - -	10 ,,	10 ,,
Tan and sawdust - -	46 ,,	50 ,,

To each of the above receipts 30 quarts of water are added, and the compound mixed, dried, and sifted. The proportions may be varied, greater strength being given by increasing the amount of chlorate of potash and diminishing that of the nitrate of soda. (Spec. No. 1,796, 17.6.62, D. 612.)

473. **Kieselguhr,** a silicious infusorial earth, found in Hanover and other places, and largely used in the manufacture of dynamite. It has the property of absorbing from three to four times its own weight of nitro-glycerine.

474. **R.** V_1. **Kinetite** (T. Petry and O. Fallenstein) consists of nitro-benzol thickened or gelatinised by the addition of some collodion-cotton incorporated with finely ground chlorate of potash and precipitated sulphide of antimony. It requires a very high temperature, comparatively, to ignite it, and cannot under ordinary circumstances, when

Kinetite—*continued.*

unconfined be exploded by the application of heat. It is but little affected by immersion in water unless it be prolonged, when the chlorate dissolves out, leaving a practically inexplosive residue.* But it is, unfortunately, extremely sensitive to combined friction and percussion and is readily ignited by a glancing blow of wood upon wood. It is also deficient in chemical stability, and has been known to ignite spontaneously both in the laboratory and in a magazine.

It is an orange coloured, plastic mass, with the characteristic strong smell of nitro-benzol.

A variety of it contains nitrate of potash in lieu of sulphide of antimony, but this, though slightly less sensitive than that described above, is still dangerously so to combined friction and percussion.

The proportions found in a sample of the first variety were :—

Nitro-benzol	19·4
Chlorate of potash	76·9
Sulphide of antimony } Nitro-cotton }	3·7†

(Spec. No. 10,936, 6.7.84.) *See* also DULITZ and FALLENSTEIN.

475. R. V_1. **Kitchen** submitted a sample of an explosive named Cycene, consisting of :—

Chlorate of potash	8 parts.
Coal dust	3 ,,
Resin or sulphur	1 part.

It proved to be sensitive to friction or percussion after keeping. (Spec. No. 11,102, 1889.)

476. A.‡ IV_2. **Kitchen** afterwards submitted a sample consisting of :—

Chlorate of potash	3 parts.
Nitrate of potash	7 ,,
Sugar	7 ,,
Coal dust and paraffin oil	1 part.

This proved less sensitive, and was favourably reported on.

* If, however, it be exposed to moist and dry air alternately, the chlorate crystallises out on to the surface and renders the explosive very sensitive.

† A full account of this explosive is given in a lecture reported in the "Journal of the Society of Chemical Industry" for January 1887.

‡ No application for license has been received.

476 ᵃ. **K.M.P. Powder.** *See* PLASTOMENITE.

477. II. **Knab** proposed to remove the hydrometric tendency of nitrate of soda by mixing the dry salt with about 4 per cent. of oil. The French Explosives Committee reported unfavourably on the proposal. (P. & S. 500.)

478. V_1. **Knaffl's Powder** consists of :—

Chlorate of potash	46 parts.
Nitrate of potash	26 „
Sulphur	15 „
Ulmate of ammonia	10 „

Ulmate of ammonia is a brown matter obtained by exposing a mixed cotton and wool fabric to the action of superheated steam. The cotton is untouched, but the wool forms a brown matter easily removed by a beating machine. It is usually sold for manure.

As might be expected from its composition, the above powder is very sensitive. (D., p. 612.)

479. V_1. **Köhler's Powder** consists of :—

Chlorate of potash	70 per cent.
Sulphur	20 „
Charcoal	10 „

Obviously a dangerously sensitive powder. (Spec. No. 1,622, 10.6.57.)

480. IV_1. **Kolbe** has patented an explosive for use in coal mines, for which he claims that it will not ignite fire-damp or coal dust. It is composed of an intimate mixture of nitro-glycerine with not more than 20 per cent. of carbonate, oxalate, or chlorate of ammonia. (Fr. Spec. No. 189,588, 26.3.88.)

481. III_2. **Kolf** has patented a smokeless powder, which he claims to be free from danger in handling. He employs nitrated hydro-carbons (plant or vegetable refuse such as malt grains, residue from the steeping vats of a brewery, &c.), the resulting nitro-compound being afterwards :—

(*a.*) Sulphuretted under pressure by treatment with sulphuretted hydrogen, sulphydrates, and poly-sulphides.

(*b.*) Charged under pressure with some oxidising material previously dissolved.

(*c.*) Impregnated by compression with a dinitro-hydro-carbon or dinitro-cellulose to give it the required form.

(Spec. No. 8,811, 7.6.90. Fr. Spec. No. 206,198, 7.6.90.)

482. III $_1$. **Kolf** has also patented a simplification of the above, which consists in mixing the nitrated hydrocarbon with nitrates and nitro-sugar, nitro-treacle, or nitro-glycerine. The mass is warmed and pressed to the required shape. (Spec. No. 22,739, 10.12.92.)

Application has been made for license for this explosive under the name of Kolf's Blasting Powder. The following proportions are given :—

Nitro-carbon (? nitro-hydrocarbon)	50	per cent.
Nitro-sugar - - -	38	„
Nitro-glycerine - - - -	8	„
Saltpetre - - - -	2	„
Aniline - - - - - -	2	„

Samples have just been received for examination (December, 1894).

483. **A.** III $_2$. **Kolf's Powder** is defined as consisting of nitro-cotton one part, nitro-lignin and nitro-starch together one part, gelatinised by means of a suitable solvent. To this mixture not more more than 2 per cent. of nitrate of potash and 0·5 per cent. of sulphur may be added. The powder thus submitted consisted of thin greenish-brown plates, nearly square. It is, no doubt, manufactured under the patent referred to in (481).

484. III $_1$. **Kolner Dynamite Factory** has patented :—

(1.) The application of solutions of nitrate of ammonia in liquid ammonia to replace or regulate the amount of nitro-glycerine entering into the composition of an explosive.

(2.) The manufacture of the said nitrate of ammonia by means of the nitric acid vapour given off in the nitro-glycerine manufacture by introducing it into a carbon cylinder in which it meets with ammonia vapour. The resulting salt is afterwards crystallised out.

(3.) The application of oleate or margarate of alumina as a means of impregnation, with a view to regulating the amount of moisture necessary to bring the cartridges to the required consistency.

(Fr. Spec. No. 169,406, 6.6.85.)

485. II. **König** mixes nitrate of ammonia with resin, paraffin, or wax. (Spec. No. 3,024, 12.2.94.) *See* WESTFALITE.

486. **Köppel.** *See* VULCANITE.

487. IV$_1$. **Kraft** is a sort of dynamite introduced by C. J. Bjorkmann, and consists of:—

Nitro-glycerine - - - - -	62 parts.
Chlorate of potash - - -	19 „
Nitrate „ „ - - - - -	17 „
Ground cork - - - -	14 „

488. **Krebs.** *See* LITHOFRACTEUR.

489. III$_1$. **Krümmel** has proposed a dynamite consisting of:—

—	1.	2.
Nitro-glycerine - - - - - -	40 to 50 parts	30 to 35 parts.
Nitrated wood pulp - - -	10 „	60 „
Kieselguhr - - - - - - -	40 „	5 „

These two mixtures form brownish materials. The less powerful composition is intended for use in coal mines or for other soft material. (D. 726.)

490. III$_1$. **Kübin** and **Sierch** have invented a dynamite for fiery mines containing 20 per cent. to 50 per cent. of chloride or sulphate of ammonia or of a mixture of these two salts. (Spec. No. 3,759, 10.3.88.) *See* WETTER.

491. II. **Kübin** has patented the following mixture:—

Nitrate of ammonia - - -	95 per cent.	to 75 per cent.
Nitrate of aniline or toluidine -	5 „	to 25 „

(Spec. No. 11,502, 7.4.94.)

492. III$_1$. **Kuhnt** and **Diessler** mix nitro-glycerine with 60 per cent. of carbonate or chloride of ammonium, and claim that the temperature of the gases evolved on explosion is so low as to prevent flame. (Spec. No. 5,949, 21.4.88.) *See* WETTER.

493. II. **Kup's Powder** contains 80 per cent. of nitrate of baryta, with sulphur and charcoal. (D. 610.)

494. III$_2$. **Lafaye** has patented the substitution of purified wood meal for cotton in the manufacture of nitro-cellulose, with the object of reducing the cost. (Fr. Spec. No. 192,369, 29.10.88.)

495. **Laligant.** *See* SANLAVILLE and LALIGANT.

496. III$_2$. **Lambotte** makes the three following mixtures:—

First.

Sulphuric acid (*s.g.*, 1·846)	-	53·20 parts.
Fuming nitric acid	-	26·60 „

Second.

Saccharine matter in solution	-	10·040 „
Bisulphide of carbon	-	0·050 „
Oxide of lead	-	0·025 „

Third.

Sawdust or bran	-	6·740 „
Sulphate of baryta	-	3·345 „

The first mixture must be cooled to at least 68° F., the second must be heated to 266° F. (*sic*) at least if required immediately, or must be left cool for seven days at least before being taken into use.

The two first mixtures are placed in intimate contact, little by little, and the whole is then thrown into six or seven times its volume of cold water. The precipitate formed is then thoroughly washed with water which may be heated to boiling point. This substance is then incorporated with the third mixture so as to form a homogeneous paste.

The proportions of bisulphide of carbon and oxide of lead may be varied, as also may the ingredients of the third mixture. (Fr. Spec. No. 152,285, 24.11.82.)

497. **Lamm (C.).** *See* BELLITE and NITROLITE.

498. III$_2$. **Lamm (C.)** proposes the following mixtures:—

—	A.	B.	C.	D.
Dinitro-benzol	1·0	1·0	1·0	1·0
Nitrate of ammonia	1·9	—	—	—
„ „ potash	—	0·96	—	—
„ „ baryta	—	—	1·24	—
„ „ soda	—	—	—	0·81

See BELLITE, &c. (A. & E. I. 110.)

499. **Lamm** has patented the use of palm wax or carnauba wax, either alone or in combination with substances having a lower melting point, as a protection for explosives, particularly for those containing hygroscopic salts. (Fr. Spec. No. 192,974, 14.9.88.)

500. IV $_{1 \& 2}$. V $_{1}$. **Landener** has patented the treatment of chlorates and perchlorates with fatty matters, hydrocarburetted or nitrated. He gives the following formulæ:—

1.

Chlorate of potash	5 to 10 parts.
Sulphur	½ ,,
Dinitro-naphthaline	5 ,,
Tar	5 ,,

2.

Chlorate of potash	5 to 10 parts.
Nitro-cellulose	20 ,,
Cocoa-nut oil	10 ,,
Tar	10 ,,

3.

Chlorate of potash	5 to 10 parts.
Nitro-glycerine	10 ,,
Wood tar	25 ,,

He has also patented the use of special detonating composition. (Spec. No. 19,267, 7.11.91. Fr. Spec. No 216,053, 11.9.91, 1.10.91, 6.2.92.)

501. III $_{1}$. **Landsdorf** has proposed:—

Nitro-glycerine	75 per cent.
Urate of ammonia	5 ,,
Kieselguhr	20 ,,

502. II. **Landsdorf** has also proposed the following mixture:—

Nitrate of potash	73 per cent.
Urate of ammonia	9 ,,
Sulphur	9 ,,
Charcoal	9 ,,

503. III $_{1 \& 2}$. **Lanfrey's Powder** consists of straw treated with the usual acids to produce a nitro-cellulose, which is afterwards impregnated with a solution containing saltpetre, hardwood, charcoal, and dextrine. It was proposed

Lanfrey's Powder—*continued.*

also to impregnate this nitro-straw with nitro-glycerine to form a "straw dynamite." The silica present in the straw is said to give stability, though on what grounds it is not quite easy to see. (Spec. No. 3,119, 7.8 78.)

504. **Lange.** *See* SELWIG and LANGE.

505. III_2. **Lannoy Powder.** This is a white powder composed of :—

Nitrate of soda	-	65 parts.
Sulphur	-	13 „
Wood, sawdust, or bran nitrated	-	22 „

It is stated to ignite with difficulty and to burn slowly in air. It gives out strong sickening fumes. This powder was originally called Lithofracteur. (D., p. 669.)

506. VI_3. **Lauer Detonator,** which is used in Austrian coal-mines, is a frictional detonator resembling in its action a Christmas cracker. It is fired with a string. (A. & E. II. 174.)

507. **Lebbrecht.** *See* WETTEREN.

508. **Le Bricquir.** *See* ESPIR.

509. III_2. **Lederit** (Jno. Waffen) is made in Austria, and consists of :—

Nitrate of potash	-	45 per cent.
Sulphur	-	15 „
Red lead	-	20 „
Picric acid	-	2 „
Leather cuttings	-	18 „

(O.G.)

510. **Lee.** *See* L. P. POWDER.

511. IV_2. **Le Maréchal** has patented the manufacture of a powder by mixing stearic or palmitic acid with chlorate of potash, soda, or ammonia. These ingredients are mixed hot, the chlorates being in a finely-divided condition. The resulting product is pulverised and then pressed into a homogeneous mass with or without the addition of ground charcoal. The powder is then granulated. (Fr. Spec. No. 167,943, 25.3.85.)

512. III $_2$. **Lénite.** A mixture of picric acid and collodion. (P. & S. 530.)

513. III $_1$. **Leonard** has patented as a smokeless powder the following :—

Nitro-glycerine	-	150 parts.
Gun-cotton	-	50 „
Lycopodium	-	10 „
Urea	-	4 „

The ingredients are incorporated with the aid of acetone. (Spec. No. 20,066, 25.11.93.)

514. III $_1$. **Lewin** has patented an explosive called Sandholite. It is composed of :—

Nitro-glycerine.
Nitro-cotton.
Nitro-sugar (cane).
Rye-meal.
Nitrate of soda.
Paraffin.
Tar.

He has also patented the use of nitrated cane sugar or cane-sugar residue as an explosive alone or in mixture. (Fr. Spec. No. 185,956, 20.9.87.)

515. **Lewin.** *See* FORCITE.

516. III $_2$. **Liardet** dissolves picric acid in half its weight of boiling glycerine, and adds a proportion of ground cedar or other wood, and nitrate of potash. (Spec. No. 12,427, 6.8.89.)

516[a]. **Liardet.** *See* ACME POWDER.

517. III $_1$. **Liebert** adds 3 to 5 per cent. of iso-amyl-nitrate to nitro-glycerine, or nitrates an emulsion of glycerine and iso-amyl-nitrate, or iso-amyl-alcohol. He claims that this will not freeze at — 35°, and that it is less sensitive and more powerful than nitro-glycerine. (Spec. No. 5,503, 30.3.89.)

518. III $_1$. **Liebert** has also patented a method of manufacturing nitro-glycerine which consists of adding sulphate of iron or nitrate of ammonia to the acids employed in the manufacture. (Fr. Spec. No. 198,726, 4.6.89.)

519. II. **Liesch's Powder,** also called Petralit, consists of nitre, sulphur, wood-pulp, and coke-dust. It is made in Hungary. (O. G.)

520. VII_1. **Lightning Paper** is rather a toy firework than an explosive. It consists of thin sheets of paper treated with the usual acids and is a mixture of trinitro-cellulose and lower forms. It is impregnated with various metallic salts to give a coloured flame.

521. III_1. **Lignin-dynamite** is a generic name for mixtures of nitro-glycerine with sawdust or wood-pulp, whether nitrated or not. In some forms nitrates are also added.

522. III_1. **Lignose,** a synonym for Lignin-dynamite.

523. III_2. **Lignose** is also a name given to nitrated wood as made in a Prussian factory.

524. V_2. **Limparicht** brought to light, in 1888, the following explosives, which appear to have only a theoretical interest:—

(1.) Meta-triazo-benzol-sulphate of baryta, a body which crystallises in fine colourless needles, and explodes at 130° C.

(2.) Triazo-nitro-benzol-sulphate of potash, which crystallises in brilliant brown scales and is very unstable. It explodes at 130° C.

(3.) Sulpho-diazo-triazo-benzol acid, which takes the form of orange-red crystals, changing to deep blue on exposure to air. It is very sensitive to heat and shock.

(4.) Sulpho-diazo-dibromo-benzol acid yellow crystals even more sensitive than (3).

(5.) Triazo-dibromo-benzol-sulphate of baryta crystallises in light red scales.

(6.) Hydrazino-benzol-disulphuric acid, crystallising in brilliant rhombs.

(7.) Triazo-benzol-disulphate of baryta, which crystallises into yellowish scales, decomposing readily at ordinary temperature. (Salvati.)

525. **Lithoclastite.** *See* Roca.

526. **A.** III$_1$. **Lithofracteur** is a sort of Dynamite No. 2. As imported into England it consists of not more than 55 parts of nitro-glycerine mixed with 45 parts of a pulverised preparation consisting of one part of charcoal, bran, and sawdust (singly or in combination), 3½ parts of kieselguhr, 2½ parts of nitrate of baryta and bicarbonate of soda (or either of them), and ½ part of sulphur and manganese (or either of them). Analyses have given :—

—	A.	B.
	Per Cent.	Per Cent.
Nitro-glycerine	52	70
Kieselguhr and sand	30	23
Powdered coal	12	2
Nitrate of soda	4	—
„ „ baryta	—	5
Sulphur	2	—

It is inferior in explosive power to Dynamite No. 1, and is said to be more sensitive to heat. To all intents it is a mixture of ordinary dynamite with a crude sort of gunpowder. Though on the list of authorised explosives it is now rarely used in this country.

527. III$_1$. **Lithofracteur No. 2.** This explosive is authorised in the Colony of Victoria. It is composed of nitro-glycerine 64 per cent., absorbed in a mixture of nitrates of potash and baryta, charcoal, wood-meal, manganese, and carbonate of magnesia.

528. **Lithofracteur.** *See* Saxifragine, Rendrock, Lannoy.

529 **A.** III$_2$. **Lithotrite,** patented by Antheunis, consists of the following ingredients mixed in a revolving drum :—

Nitrated mahogany wood-pulp	8 per cent.
Nitrate of potash	50 „
Nitrate of soda	16 „
Wood charcoal	1·5 „
Sublimed sulphur	18 „
Ferro-cyanide of potash	3 „
Picrate of ammonia	3·5 „

It is a fine grey powder, sometimes made up into compressed cartridges. As submitted in 1888, this explosive consisted only of charcoal, sulphur, saltpetre,

Lithotrite—*continued.*
sawdust, and a little nitrate of soda. (Spec. No. 783, 1887. Fr. Spec. No. 166,874, 7.2.85.)

530. **Liverpool Cotton Powder.** *See* POTENTITE.

531. **Lloyd.** *See* GALLAHER.

532. II. **Lobb's Powder.** In this powder sawdust is substituted for charcoal and sulphur in gunpowder, and lime is added. The object of the sawdust is to reduce the amount of smoke. The lime is added to counteract the deliquescent property of nitrates of soda or potash in gunpowder, especially in underground or damp situations. (Spec. No. 1,861, 25.10.61.)

533. **Lom de Berg (de).** *See* MAGNIER.

534. **Lorrain.** *See* SCHNEBELITE.

535. V_1. **Lovelace's Detonating Composition** consists of picrate of mercury, picric acid, and chlorate of potash.

536. V_1. **Lovelace's Powder** consists of picric acid, chlorate of potash, and charcoal.

537. III_1. **L. P. Powder,** a name under which a Belgian smokeless powder was submitted. It is a powder in the form of small black cubes, smelling of butyric ether ($C_4H_5 . C_8H_7O_4$). Its composition was—

Nitro-glycerine . . .	32·4 per cent.
Nitro-cellulose	60·2 ,,
Charcoal	7·4 ,,

538. III_1. **Lundholm** and **Sayers** have patented a method of mixing nitro-glycerine and nitro-cellulose in water for the subsequent manufacture of ballistite, cordite, &c. (Specs. Nos. 10,376, 26.6.89, and 6,448, 4.4.92.)

539. III., IV. **Lundholm** and **Sayers** have also proposed to nitrate hydro- and oxy-cellulose. The resulting product may be mixed with camphor, nitro-glycerine, nitrates, chlorates, &c. (Spec. No. 6,399, 15.4.89.) *See* HYDRO-NITRO-CELLULOSE.

540. III_2. **Lyddite.** An explosive for use in shells, with which trials have been carried out by the War Office. Its principal constituent is picric acid, and in this respect it resembles Melinite (*q.v.*).

541. III$_2$. **MacGavin** has proposed to saturate sawdust with a solution of picrate of potash. The sawdust is then dried and mixed with nitrates of potash and soda and with sulphur. (Spec. No. 9,433, 7.6,89.)

542. II., IV$_2$. **Macintosh** proposed to mix gunpowder and other explosive compounds with india-rubber or gutta-percha in a state of solution, and to spread the mixture on cloth, which becomes inflammable, burning rapidly if mixed with chlorate or nitrate of potash, and slowly if mixed with steel filings or sulphur. The cloth can be cut into strips and used in connexion with incendiary projectiles. (Spec. No. 404, 11.2.57.)

543. III$_2$. **Mackie,** in his patent taken out in conjunction with Faure for a process of manufacturing gun-cotton, also proposed a blasting powder consisting of nitrated gun-cotton (or similar substance) mixed with resin, shellac, ozokerit, collodion, glycerine, charcoal, or soot. As an instance a mixture is given of five to ten parts gun-cotton with two parts of resin and one of nitre. If glycerine or ozokerit be used a plastic compound is obtained. (Spec. No. 1,830, 20.5.73.)

In conjunction with Faure and Trench he subsequently patented the admixture of nitrate of baryta with gun-cotton (*see* TONITE). (Spec. No. 3,612, 20.10.74.)

Afterwards the same persons patented the use of esparto grass, hemp, flax, straw, hay, agave, and yucca fibre, and other vegetable substances as substitutes for cotton. (Spec. No. 2,742, 4.7.76.)

543a. **A.** I. **MacNab (J.)** proposes to insert a glass tube containing aqueous solution of ammonia into gunpowder cartridges, with the view of cooling the flame, and rendering the cartridges safe for use in fiery mines. Other advantages are also claimed.

544. **A.** VII$_2$. **Magic Candle Pin Crackers** consist of minute quantities of fulminate of silver attached to pins. They are intended to be stuck into a candle and to fire when the pin becomes heated.

545. **Magnesia Powder.** *See* HERCULES POWDER NO. 1.

545a. III $_2$. **Magnier, De Lom de Berg,** and **Vielliard** have patented the following:—

(1.) The nitration of the phenols and homophenols as well as the mixed products, quinone and phenol-alcohols.

(2.) The treatment of the trinitro products thus obtained, with ammonia, soda, &c., or with their carbonates, with a view to transforming them into salts of these bases; and their subsequent mixture with nitrates of ammonia, potash, soda, and baryta. (Fr Spec. No. 213,976, 8.6.91.) *See* MANUELITE and PICRIC POWDER.

546. VI $_2$. **Maissin** has patented:—

(1.) The employment of nitro-cotton in grains to form a fuze capable of transmitting fire or detonation to a great distance at the rate of several miles a second.

(2.) The employment of a special mill for preparing the nitro-cotton.

(3.) The preparation of the uncovered core of the fuze any required length. (Fr. Spec. No. 190,073, 22.4.88.) *See* DETONATING FUZE.

547. III $_2$. **Maïzite,** a blasting explosive proposed by Pesci and Zini in 1886. The composition is:—

——	No. 1.	No. 2.
	Per Cent.	Per Cent.
Picrate of ammonia - - - -	60	28
Nitrate of ammonia - - -	40	72

(P. & S. 562. *See* PICRIC POWDER.)

548. I. **Mammoth Powder,** an American black powder, of large irregular grain, formerly employed for heavy guns. (D. 344.)

549. III $_2$. **Manuelite.** This explosive is composed of 100 parts of picric acid, combined with 20 parts of carbonate of ammonia, or with 45 parts of carbonate of soda.

Manuelite—*continued.*

The resulting salts of ammonia and soda are then employed in the following mixtures:—

1.

Ammonia salts	28	per cent.
Nitrate of ammonia	72	,,

2.

Ammonia salts	40	,,
Nitrate of potash	60	,,

3.

Soda salts	40	,,
Nitrate of potash	60	,,

The resinous hydrocarbons which separate during the nitration of nitro-derivatives are added to give the explosive a plastic consistency.

Practically these explosives are simply mixtures of picrates and nitrates, the advantage of adding the resinous hydro-carbon being very doubtful. (Fr. Spec. No. 213,976, 8.6.91.)

See MAGNIER, PICRIC POWDER, &c.

550. **Marechal.** *See* LE MARECHAL.

551. **Mastodon Powder.** *See* MAMMOTH POWDER.

552. **A. III$_1$. Matagnite.** The two following explosives are now licensed thus:—

Blasting Matagnite, consisting of nitro-cellulose carefully washed and purified, combined with thoroughly purified nitro-glycerine and thoroughly purified nitro-benzol, or either of them, in such proportions that the whole shall be of such a character and consistency as not to be liable to liquefaction or exudation.

Matagnite Gelatine, consisting of a thoroughly purified nitro-glycerine, and thoroughly purified nitro-benzol, or either of them, thickened by being combined with nitro-cellulose carefully washed and purified, and mixed or incorporated with wood meal for the bleaching or purifying of which no chemical agents have been used, or if used have been absolutely removed, and nitrate of potassium, or such other nitrate as may for the time being be sanctioned by a Secretary of State, in such proportions that the whole shall be of such character and

Matagnite—*continued.*

consistency as not to be liable to liquefaction or exudation.

A sample of Matagnite Gelatine gave:—

Nitro-glycerine	68·00	per cent.
Nitro-cotton and wood meal	11·40	„
Nitrate of potash	20·60	„

553. III$_1$. **Mataziette** was manufactured at Fabry near Geneva. It consisted of a mixture of about 40 per cent. of nitroglycerine with sand, ochre, coarsely pounded charcoal, and resinous matter. In October 1877, twelve casks, containing in all about three tons of this material, were seized by the French douaniers at Pontarlier, an attempt having been made to pass it as manure. It was removed to the fort of Larmont for safe custody. The casks cracked and allowed the contents to escape, and after three months it was decided to re-pack the explosives into boxes lined with sawdust. The work was half done when an explosion took place killing six and wounding four men, and doing considerable amount of structural damage. The accident was due to rash and rough handling of the explosive by inexperienced men.

554. **Matteen.** *See* PYROLITHE.

554[a]. V$_1$. **Maurette** proposes:—

Nitrate of potash	640	parts
Charcoal	50	„
Wood ashes	55	„
Sulphur	25	„
Chlorate of potash	5	„

555. II. **Maxim** proposed a mixture of—

Nitrate of potash	74·18	parts.
Paraffin	11·42	„
Sulphur	10·40	„

He mixes one part of this with three parts of gunpowder to obtain a plastic cartridge. (Spec. No. 6,926, 8.6.85.)

556. III$_2$., IV$_2$. **Maxim** treats gun-cotton with solvents such as acetone or acetone alcohol and ether in an exhausted vessel, so as to get rid of air-bubles, and obtains a very hard material. He also claims the addition of chlorate of potash or other oxidising salt to collodion cotton. (Spec. No. 16,213, 8.12.88, Fr. Spec. No. 194,792, 15.12.88.)

557. III$_1$. **Maxim** also proposes to add oil (preferably castor oil) to compounds of dissolved gun-cotton, nitro-glycerine, &c., to produce a slow burning explosive for small arms. Suitable proportions are 2 to 5 per cent. castor oil, 10 to 16 per cent. nitro-glycerine and the remainder gun-cotton. (Spec. No. 4,477, 14.4.89.)

557[a]. V$_1$. **Maxim** proposes as a detonating composition which can be fired with safety in shells the following:—

Nitro-glycerine - 75 to 85 parts	} 25 to 15 per cent.	
Gun-cotton - 25 to 15 ,,		
Fulminate of mercury -	- 75 to 85 ,,	

(Spec. No. 18682, 2.10.94.)

558. III$_2$. **Maximite** is an explosive invented by Mr. Hudson Maxim, and subjected to public trials in America. The composition is not given, but the newspaper accounts state that it is a nitro-compound with a gun-cotton base. It is claimed that the explosive is as powerful as blasting gelatine, and is unfreezable.

559. I. **Maxwell** simply diminished the amount of saltpetre in gunpowder "4 per cent., more or less," and used spirit, or spirit and water, instead of water alone in liquoring. (Spec. No. 1,066, 27.4.60.)

560. **Médail.** *See* BENGALINE.

561. R. III$_1$. **Meganite** (Schückher & Co.) is of three kinds:—

—	I.	II.	III.
Nitro-glycerine - -	60 per cent.	38 per cent.	7 per cent.
Nitro-lignin - -	10 ,,	6 ,,	9 ,,
Nitrated vegetable ivory -	10 ,,	6 ,,	9 ,,
"Added powder" - -	20 ,,	50 ,,	70 ,,

The "added powder" for No. 1 is nitrate of soda; for No. 2, 75 parts of nitrate of soda to 1 of soda and 24 of wood meal; for No. 3, as for No. 2, except that rye meal takes the place of wood meal. *See also* ORIASITE.

562. III$_1$. **Melanite,** a variety of blasting gelatine, proposed by Faille, and containing 13 per cent. to 17 per cent. of nitro-cellulose and 87 per cent. to 83 per cent. of nitro-glycerine.

563. III$_2$. **Melinite** is an explosive which has recently been experimented with on a large scale by the French Government. Its composition is kept secret, but picric acid is supposed to form its main constituent, and it very possibly much resembles the explosive patented by Turpin, and mentioned in the introduction. *See* p. xxxix.

There is some reason to believe that nitro-benzol or a similar material is employed as well. A portion of the explosive, as used to fill shells, is said to be termed Crésilite, probably dinitro-cresol ($C_7H_5(NO_2)_2OH$) with which two-thirds of the shell is filled, and the remainder with Melinite. Cresylic acid (C_7H_8O) corresponds to carbolic acid, but contains the (hypothetical) radical cresyl C_7H_7 in lieu of phenyl (C_6H_5).

Mr. Guttman states that Melinite was originally picric acid embedded in collodion. Later, only fused picric acid was used with a gun-cotton primer. Nitro-cresol was sometimes added, but subsequently abandoned.

564. IV$_2$. **Melland's Paper Powder** consists of unsized paper dipped into a boiling solution of the ingredients given below. The paper is then rolled into cartridges, and dried at 212° F. To protect the cartridges from damp they are coated with a solution of one part of xyloidine (nitro-starch) in three parts of acetic acid. The ingredients are as follows:—

Chlorate of potash	9	parts.
Nitrate of potash	$4\frac{1}{2}$	„
Ferrocyanide of potash	$3\frac{1}{4}$	„
Charcoal	$3\frac{1}{4}$	„
Starch	$\frac{1}{21}$	„
Chromate of potash	$\frac{1}{16}$	„

(D., p. 614).

This was patented in England by Reichen. (Spec. No. 2,266, 2.9.65.)

565. IV$_2$. **Melville's Powders** consist of:—

—	I.	II.	III.
Chlorate of potash	2 parts.	5 parts.	1 part.
Red orpiment	1 „	2 „	— „
Prussiate of potash	— „	1 „	1 „

The ingredients are well mixed, damped, and moulded into cartridges for firearms or for mining purposes, or charging shells. (Spec. No. 13,215, 6.10.50.)

566. III_1. **Mendoça-Corteso Powder.** This is a Portuguese smokeless powder of similar composition to Cordite. (*q.v.*)

567. III_1. **Merino** proposes the protection of nitrates or chlorates by means of an envelope of melted hydrocarbon. He suggests the following composition:—

Saltpetre	73	per cent.
Sulphur	7	„
Caoutchouc or pitch	3	„
Resin or tar	1	„
Anthracite	10	„
Nitro-glycerine	6	„

(Fr. Spec. No. 151,960, 7.11.82.)

568. **Mertz.** *See* Rosenboom and Mertz.

569. **Metallic Safety Fuze.** *See* SAFETY FUZE.

570. III_1. **Metalline Nitroleum** is a mixture of nitro-glycerine with pulverised red lead, with or without plaster of Paris.

571. **Meurling.** *See* NORDENFELT.

572. V_1. **Meyer** and **Moritz** proposed a mixture of :—

Fine gunpowder	20	parts.
Antimony	1	„
Saltpetre	2	„

One part of this mixture and five parts of fulminate of mercury were to be mixed with half a part of Roman cement, and moistened with gum to form a paste. When dry it was to be waterproofed by means of grease. It is doubtful if this curious mixture was ever really tried. (Spec. No. 515, 23.2.65.)

573. I. **M. G[1]. Powder** was introduced into the Service in 1882 for use in machine guns. It is highly glazed and somewhat angular in shape, the size of grain being from 7 to 14 mesh. It is so made that it can be compressed into a pellet. Its density is 1·75.

574. III_1. **Mica Powder** is a mixture of nitro-glycerine with finely divided scales of mica, and is due to Mowbray. It is claimed that the mica scales, which replaced the thin flakes of glass originally tried, act as carriers rather than absorbents of the liquid explosive. The principle involved is the same as that enunciated by Judson, *q.v.*, namely, that a proportion of nitro-glycerine is practically

Mica Powder—*continued.*
lost in ordinary absorbents such as kieselguhr. But the mica scale will only retain 52 per cent. of nitro-glycerine as compared with the 75 per cent. of ordinary dynamite. The mica powder is said to be quicker in action than ordinary dynamite. (T., 88.)

575. IV_2. **Michalowski's Miners' Powder** consists of :—

Chlorate of potash - - -	50 parts.
Binoxide of manganese - - -	5 ,,
Finely pulverised organic matter -	45 ,,

The last ingredient is wood sawdust, tan, bran, &c., usually the latter. The explosive much resembles Asphaline, *q.v.* (M. XIII., p. 244. Fr. Spec. No. 148,681, 1.5.82.)

576. V_1. **Millbank** proposed as a composition for "caps primers, and cartridges," two detonating mixtures composed of :—

—	1.	2.
	Parts.	Parts.
Chlorate of potash - - - -	80	20
Prussiate of potash - - -	—	10
Amorphous phosphorus - - -	4½	1
Charcoal - - - - -	35	—

He speaks of them as reliable and safe! (T., p. 103.)

577. II. **Miller's Powder** consits of two compounds, which are harmless apart, and become explosive only when mixed.

No. 1 compound contains :—

Nitrate of soda - - - - -	35 parts.
Saltpetre - - - -	25 ,,
Starch - - - - - - -	2 ,,

No. 2 compound contains :—

Bichromate of potash - - -	3 parts.
Sulphur - - - -	13 ,,
Charcoal - - - - - - -	12 ,,

The proportions to form an explosive compound are 18 parts of No. 1 mixed with 7 parts of No. 2. (T., p. 106.)

578. **Millot.** *See* GIRARD.

579. **Mindeleff.** *See* TERRORITE.

580. III$_1$. **Miner's Friend Powder** is a lignin-dynamite containing nitrate of sodium.

581. **Miner's Powder.** *See* MICHALOWSKI.

582. III$_1$. **Miner's Powder Company Dynamite** is a No. 2 dynamite, practically identical with Vulcan Powder.

583. **Miners' Safety Explosive** is now called Ammonite, *q.v.*

584. **A.** VI$_2$. **Miner's Squibs** are identical, practically, with Hunter's Mining Fuzes. The invention was patented by Daddow. (Spec. No. 1,558, 2.5.74.) They are largely used in the Cleveland iron district.

585. III$_2$. **M. N. Smokeless Powder** (Maxim Nordenfelt), an American smokeless powder, consisting of gun-cotton gelatinised in acetic ether. *See* MAXIM (556).

586. III$_1$. **Monakay's Explosive.** To every pound of a mixture composed of equal weights of ashes, lampblack, earth, nitrate of soda, and borax, half a gill of kerosene oil is added. To this compound nitro-glycerine is added in quantities varying according to the strength desired. It is claimed that the kerosene, a fluid hydrocarbon, dilutes and modifies the nitro-glycerine so as to diminish the risks of accident, and that it acts beneficially in adding to the strength of the compound when exploded. (T., p. 106.)

587. IV$_2$. **Monnier's Powder** consists of:—

Chlorate of potash	71 per cent.
Sugar	16 „
Charcoal	6 „
Coal tar	7 „

The chlorate is dissolved in 250 parts of water, and freedom from danger is claimed for the wet process of manufacture. (T., 106.)

588. **Mononitro-benzol,-naphthaline, &c.** *See* NITRO-BENZOL, &C.

589. III$_2$. **Montravel** adds dinitro-benzol to the ingredients of gunpowder. The mixture is heated to 140°, when the dinitro-benzol and sulphur melt and form a waterproof protecting film over the grains. (Spec. No. 5,031, 22.4.89.) *See* also FORTISINE and WIENER.

590. **Moritz.** *See* MEYER and VULCANITE.

591. III$_1$. **Morse's Explosive** consists of nitro-glycerine and resin dissolved in a common solvent. The latter is evaporated and the explosive compound formed into a hard, dry, solid, granulated, or pulverised mass. (T., p. 107.)

592. **Moschek.** *See* DYNAMOÏTE.

593. R. III$_1$. **Mosenthal, Salamon,** and **Hood** propose to substitute for the whole or part of the kieselguhr in dynamite, Weldon mud, which consists principally of binoxide of manganese. (Spec. No. 13,038, 31.7.91.)

594. **Mowbray.** *See* MICA.

595. III$_1$. **Mowbray** patented a mixture of three parts nitro-toluol to seven parts of nitro-glycerine, but states that he manufactures his nitro-toluol "similarly to nitro-glycerine." Some confusion seems here to exist between the two hydrocarbons, benzol (C_6H_6) and toluol (C_7H_8). He also gives a mixture of the two ingredients in the proportion of 1 to 3. This he states to be a very safe explosive, but being a liquid not easily congealed, if a leak should occur in the containing vessel the nitro-toluol (? nitro-benzol) is liable to evaporate, leaving the nitro-glycerine alone. (T., 101.)

596. III$_2$. **Mühlhäusen** has proposed the following:—

Nitrate of ammonia - -	75 per cent.
Nitro-jute - - - - - -	25 „

(Salvati.)

597. III$_1$. **Müller** has patented the incorporation of 15 per cent. of salts containing at least 5 molecules of water of crystallisation (such as crystals of soda, sulphate alums, borate and phosphate of soda) in such explosives as contain

Müller—*continued.*

nitro-glycerine, nitro-benzol, or other equivalent of the aromatic group of coal-tar derivatives. Such composition serves for the manufacture of plastic cartridges for use in fiery mines. (Fr. Spec. No. 185,809, 13.9.87.) *See* WETTER.

598. **Mundell.** *See* PERTUISET.

599. **Munroe.** *See* INDURITE.

600. III$_1$. **Munroe-Jewell Powder,** an American smokeless powder similar to Ballistite. (P. & S. 612.)

601. II. **Murtineddu's Powders** consist of mixtures of nitrate of soda (with or without saltpetre), with sulphur and various substances as tan, coal, sawdust, &c.

The mixture, patented in England, consists of:—

Saltpetre	100 parts.
Sulphur	100 ,,
Sawdust	50 ,,
Horsedung	50 ,,
Sea salt	10 ,,
Treacle	4 ,,

The object of adding the treacle is to give cohesion to the composition. It is claimed that "this composition " does not cause explosion upwards as with gunpowder." (D., p. 608, and Spec. No. 2,403, 14.10.56.)

602. III$_2$. **Muschamp** patented a wood pulp process, and proposed to utilise the acids used in dipping the first batch of cellulose to make weaker nitro-compounds by dipping second and third batches in them. To retain rapidity of explosion he steeped his nitro-compound in a solution of sulphate of zinc or lead, or in a solution of starch. The charges when made up were to be waterproofed with collodion or other substance. (Spec. No. 1,326, 16.5.71.)

603. III$_1$. **Neptune Powder,** an American dynamite of the No. 2 type, practically identical with Vulcan Powder, Vigorite, &c.

604. I. **Neumeyer** proposed gunpowder consisting of:—

Saltpetre	75·00	per cent.
Charcoal	18·75	„
Flowers of sulphur	6·25	„

He preferred charcoal made from birchwood, and soaked the charcoal in soda-lye. He claimed that gunpowder thus made burnt quietly and slowly when unconfined, but had equal force to ordinary powder when confined. (Spec. No. 1,636, 17.6.65.)

He afterwards proposed a mixture of:—

Saltpetre	72	per cent.
Charcoal	18	„
Flowers of sulphur	10	„

abandoned the use of soda-lye, and mixed his ingredients with 40 per cent. of water in a drum, fitted with revolving arms, for 15 minutes. The resulting compound was simply dried, and stated to be then fit for use. (Spec. No. 1,408, 13.5.67, and D., p. 601.)

605. **Newton.** *See* SAXIFRAGINE.

606. III_2. **Nico,** a fancy name under which Liardet's powder (*q.v.*) has been authorised in the Colony of Victoria.

607. V_1. **Nisser's Powders** consist of:—

Yellow prussiate of potash	1·5	per cent.
Bichromate	2·0	„
Perchlorate or chlorate of potash	10·5	„
Nitrates of soda and potash	44·5	„
Vegetable matter	6·5	„
Mineral and vegetable carbon	19·5	„
Sulphur	15·5	„

A more explosive powder for ordnance and firearms is made by using more of the potassium salts and less carbon and sulphur. (Spec. No. 1,939, 26.7.65.)

He also proposed two inexplosive compounds to be kept apart till required for use, when their mixture formed an explosive.

No. 1 *compound* consisted of nitrate of potash or of soda mixed with perchlorate or chlorate of potash, or both, in proportions varying from 5 to 35 per cent. of each of these salts, according to the required strength of the powder.

Nisser's Powders—*continued.*

No. 2 *compound* consisted of 25 to 35 per cent. of pure loaf sugar added to sublimed sulphur, and to these 8 to 10 per cent. of vegetable fibre or charcoal, or both.

The powder is stated not to ignite by friction or percussion. (Spec. No. 1,375, 27.4.68.)

He also proposed two compounds, consisting of chlorate of potash 55 to 60 parts, bitartrate of potash 45 to 55 parts, ferrocyanide of potash, and coal. In the second compound the ferrocyanide of potash was replaced by sulphur. (Spec. No. 119, 14.1.70. D. 614.)

608. III_2. **Nitramide** or **Nitramite,** a name given to Favier's explosives or ammonite as made in Spain and Russia. (H. M.)

609. III_2. **Nitramidine,** a name given by Dumas to pyroxiline prepared from paper or cardboard.

610. **Nitramidine.** *See* NITRO-STARCH.

611. **A.** III_2. **Nitrated Gun-cotton** is a general term applied to all admixtures of gun-cotton with nitrates. In this country nitrate of lead is forbidden to be used on account of the deleterious fumes produced by it on explosion.

612. IV_2. **Nitrate of Ammonia Powder** (Special) has the following composition:—

Nitrate of ammonia - - -	80 per cent.
Chlorate of potash - -	5 „
Nitro-glucose - - - -	10 „
Coal tar - - - -	5 „

This name has also been used for Amide Powder (*q.v.*). (M. XIII. 245.)

613. **Nitrate of Ammonia,** NH_4NO_3. *See* p. xxii.

614. **Nitrate of Baryta,** $Ba(NO_3)_2$. *See* p. xxii.

615. **Nitrate of Copper.** *See* AMMONIO-NITRATE OF COPPER.

616. **Nitrate of Diazo-benzol.** *See* ANILINE FULMINANTE.

617. III$_1$. **Nitrate of Ethyl,** an ordinary nitric ether ($C_2H_5NO_3$), is the result of treating alcohol with nitric acid. It is liquid, and boils at 172° F. A small quantity of the liquid can be burnt, but the vapour detonates with violence at about 284° F.

618. III$_1$. **Nitrate of Methyl** ($CH_3.NO_3$) has been proposed as a liquid explosive. It boils at 150° F. The vapour explodes at about 270° F., or in contact with flame, and the detonation is communicated to the liquid ether. This liquid caused an explosion at St. Denis, 12th November 1874, in this way. It is manufactured for dyeing purposes, and can be produced by acting on an alcoholic solution of nitrate of urea with nitric acid, or by distilling wood spirit with saltpetre and sulphuric acid.

619. **Nitrate of Lead** ($Pb(NO_3)_2$) has been used as an oxidising agent in some explosives, but on account of the poisonous nature of the fumes produced no explosive containing this ingredient is licensed in England.

620. **Nitrate of Potash,** KNO_3. *See* p. xxii.

621. **Nitrate of Soda,** $NaNO_3$. *See* p. xxii.

622. **Nitrate of Tin** merits a mention as being a possible cause of obscure explosions in powder mills. It was found at Spandau that frequent ignition of the powder took place at a certain stage of its manufacture in the Government Powder Mills. On examining the machinery it was found that where bronze pieces which were soldered were in constant contact with the moist powder the solder was much corroded, and in part entirely destroyed. In the joints a substance had collected, which when scraped out with a chisel exploded with emission of sparks. The matter was examined into and experiments were carried out, and eventually it was found out that if a thin layer of sulphur and saltpetre were placed between sheets of tin and copper foil and allowed to remain in a constantly moist condition, after a time the copper was coated with sulphide, and the tin was largely converted into an explosive basic nitrate. The action is probably this: the copper sulphide is oxidised to sulphate

Nitrate of Tin—*continued.*

which with the saltpetre forms potassium sulphate and copper nitrate. The latter, as shown by independent experiments, can unite with the tin of the solder to form the explosive basic nitrate, which, being insoluble, gradually collects in the joints and finally leads to an explosion. The substance when obtained pure is a white crystalline powder, which explodes violently with a shower of sparks when heated rapidly or subjected to percussion or friction. It is formed when a fine spray of nitric acid (*e.g*, 1·20) is thrown on a surface of tin or solder, or when tin or solder are exposed to the action of a solution of copper nitrate. (M. No. III., p. 671 *et. seq.*)

623. **Nitre.** *See* NITRATE OF POTASH, p. xxii.

624. III $_1$. **Nitresine** is a combination of nitric acid and resin proposed as an absorbent for liquid explosives. (T., p. 107.)

625. III $_2$. **Nitro-bellite** consists of a mixture of dinitro-benzol and saltpetre in the following proportions:—

—	No. 1.	No. 2.
	Per Cent.	Per Cent.
Dinitro-benzol - - - - -	45	30
Saltpetre - - - - - -	55	70

(Salvati.) *See* BELLITE.

626. **Nitro-benzoïque.** *See* VENDING.

627. III $_{1\ \&\ 2}$. **Nitro-benzol** or **Nitro-benzene.** Benzol (C_6H_6) is obtained from coal tar as a very inflammable, brilliant, colourless liquid. The action of nitric acid on this liquid produces nitro-benzol ($C_6H_5 . NO_2$) a heavy, yellow, oily liquid with a characteristic odour of bitter almonds. It is very largely used for the manufacture of aniline ($C_6H_5NH_2$), and hence for the numerous aniline dyes. By the further action of nitric acid on the mono-nitro-benzol another atom of hydrogen is displaced, and dinitro-benzol ($C_6H_4(NO_2)_2$) is obtained, which appears as

Nitro-benzol or **Nitro-benzene**—*continued.*

long, shining needles or laminæ, which melt below 100° C. and solidify into a radiated mass. *See also* INTRODUCTION, p. xl.

628. **Nitro-cellulose.** *See* p. xxxiii.

629. III_2. **Nitro-coal** is the result of the action of nitric acid on powdered coal. On account of the violent action of strong acids on the coal the nitration in the experiments made had to be carried on in stages. First nitric acid of SG. 1·40 to 1·48 was used, and after some manipulation a nitro-product was thrown down as brown powder. This was washed and dried and then treated with the strongest nitrating agents. The results obtained were not encouraging from a commercial point of view, on account of the large consumption of acid, especially when wood charcoal and coke were experimented upon. (M. No. II., p. 453.)

630. III_2. **Nitro-colle** consists of isinglass or gelatine soaked in cold water. It is then melted at a gentle heat and sufficient nitric acid is added to prevent its solidifying when cold. It is then treated with the usual acids. Another method is to place strong glue in cold water until it has absorbed the maximum amount of the latter. The mixture is solidified by the addition of nitric acid, nitrated in the usual way, and well washed.

631. III_2. **Nitro-cotton,** a generic term applied to pyroxiline made from cotton. *See* p. xxxiii.

632. $III_{1\,\&\,2}$. **Nitro-cresol.** Cresol, cresylic alcohol, or cresylic phenol (C_7H_8O), is obtained from coal or wood tar. Treated with dilute nitric acid, nitro-cresol, or nitro-cresylic acid, is deposited as a heavy oily liquid. Dinitro- and trinitro-cresol can also be produced. The latter is analogous to picric acid but contains less oxygen. *See* p. xl.

634. $III_{1\,\&\,2}$. **Nitro-cumol** or **Nitro-cumene.** Cumol (C_9H_{12}) is obtained from the products of distillation of various resins. Nitro-cumol ($C_9H_{11}NO_2$) is made by

Nitro-cumol or **Nitro-cumene**—*continued.*
dissolving cumol in strong nitric acid. It separates as a dense yellowish oil on the addition of water.

Dinitro- and trinitro-cumol are obtained by treating cumol with sulphuric and nitric acids in the ordinary way.

635. IV$_2$. **Nitro-curd** (Sjöberg). Curd is nitrated in the usual way, and an explosive made consisting of—

—	I.	II.
	Per Cent.	Per Cent.
Nitro-curd	30	3
Nitrate or oxalate of ammonium	55	55
Astral oil	10	10
Naphthalene	5	5
Chlorate of potash	—	27

(O.G.)

636. III$_2$. **Nitro-dextrin** is obtained by treating dextrin ($C_6H_{10}O_5$) with the usual acids. The resulting product resembles nitro-starch (*q.v.*).

637. III$_2$. **Nitro-flax.** Patented by Bickford and Spooner. (O. G.)

638. V$_2$. **Nitrogen Chloride** (NCl_3) is an oily liquid of a most dangerously sensitive and explosive character, and practically unmanageable for use. It is made by allowing chlorine to act on a solution of ammonium chloride.

639. V$_2$. **Nitrogen Iodide** ($NH . I_2$) is a black powder which explodes at the touch of a feather when dry. It is formed by the action of ammonia on iodine.

640. V$_2$. **Nitrogen Sulphide** (NS) is a yellow crystalline substance obtained by passing ammonia through a solution of proto-sulphide of chlorine in bisulphide of carbon. It readily explodes by percussion or when heated to about 400° F.

641. **Nitro-glucose.** *See* KEIL.

642. **Nitro-glycerine.** *See* INTRODUCTION, p. xxvi.

643. III$_1$. **Nitro-glycol.** A colourless liquid, insoluble in water, soluble in alcohol and in ether. Glycol $C_2H_4(OH)_2$ is an organic substance of the hydrocarbon class, and is transformed by the action of mixed nitric and sulphuric acids into dinitro-glycol, or "nitro-glycol." $C_4H_4(O . NO_2)_2$.

644. III $_2$. **Nitro-hemp.** Nitro-cellulose prepared from hemp. (A. & E. I. 171.)

645. III $_2$. **Nitro-jute.** Nitro-cellulose prepared from jute. (A. & E. I. 84.)

646. **Nitro-lactose.** *See* SJÖBERG.

647. III $_2$. **Nitro-lactose,** proposed as an explosive either alone or in proportion of 33 per cent. to 67 per cent. of a mixture of 45 parts nitrate of ammonium and 10 parts each of naphthalene and paraffin. It failed to pass the heat test, owing to impurity. *See also* NITRO-SACCHAROSE.

648. III $_1$. **Nitroleum,** an early name of nitro-glycerine.

649. **Nitroleum.** *See* METALLINE and PORIFERA.

650. III $_2$. **Nitro-lignin,** a generic term applied to nitro-cellulose made from woody fibre, in contradistinction to that made from cotton (nitro-cotton).

651. **Nitroline.** *See* BJORKMANN (E. A.), and BJORKMANN (C. G.).

652. III $_1$. **Nitrolite** (C. Lamm) consists of:—

Nitro-glycerine - - - -	99 to 94 parts.
Nitro-cellulose (cotton, starch, or straw nitrated) - - - -	1 ,, 6 ,,
Nitrate of ammonia, soda, or potash, and light charcoal - - -	50 ,, 150 ,,

Nitro-benzol is said also to be present. It is stated that the above is liable to exudation and is very hygroscopic.

653. IV$_1$. **Nitrolkrut.** A powder patented in Sweden by J. A. Berg in 1876. Its composition is as follows:—

Nitro-glycerine - - -	5 to 40 parts.
Chlorate of potash - -	5 ,, 50 ,,
Nitrate of potash or soda - -	25 ,, 75 ,,

The nitro-glycerine may be replaced by a nitrated hydrocarbon. (P. & S. 657.)

654. A. III$_1$. **Nitro-magnite** or **Dyna-magnite** consists essentially of nitro-glycerine absorbed in magnesia alba (a mixture of hydrocarbonates of magnesia). It was approved for manufacture in the United Kingdom in 1879, but, owing to the then existing Nobel patents which were held by a decision of the House of Lords to include all explosives of the dynamite class, no steps were taken to establish a factory. It was brought forward by Mr. E. Jones, of Caerphilly, and appeared to promise well "as the power of magnesia alba to absorb " . . . . nitro-glycerine appears even superior to " that of kieselguhr, while the volume of gas generated " by the detonation of nitro-magnite may be a little " larger, in consequence of the expulsion of carbonic acid " from the magnesium carbonate at the high temperature " to which it is exposed."*

This explosive is employed in America under the name of Hercules Powder, *q.v.*

655. III$_2$. **Nitro-mannite** ($C_6H_8(O.NO_2)_6$) is prepared by treating mannite ($C_6H_{14}O_6$) with the usual acids. Mannite or sugar of manna exists in many vegetables, but is usually prepared from manna, a saccharine juice obtained from two species of ash.

Nitro-mannite forms white needle shaped crystals, insoluble in water but soluble in ether or alcohol. Rapidly heated it ignites in about 374° F., and explodes at about 590° F. It is more susceptible to friction and percussion than nitro-glycerine. Sulphide of ammonium re-converts it into mannite. Unless pure it is liable to spontaneous decomposition.

It may be considered as the nitric ether of the hexatomic alcohol mannite. (D. 672.)

656. III$_1$. **Nitro-methane.** Methane (CH_4), the first of the paraffin series of hydrocarbons, is capable of forming the following nitro-derivatives:—

Nitro-methane	CH_3NO_2.
Dinitro-methane	$CH_2(NO_2)_2$.
Trinitro-methane (Nitroform)	$CH(NO_2)_3$.
Tetranitro-methane	$C(NO_2)_4$.

* Abel on Explosive Agents. Proc. of C.E. Vol. LXI., 1879-80. (Spec. No. 3,954, 8.10.78.)

Nitro-methane—*continued.*

These substances all take the form of heavy liquids, solidifying at from 13° to 15° C. As might be inferred from their composition they are all more or less explosive. The two last have only so far been obtained by a complicated and costly process. Tetranitro-methane contains over 65 per cent. of available oxygen besides developing heat in decomposition. It is a stable material, and only feebly explosive, though it can be fired somewhat readily by a falling weight. It dissolves paraffin and other hydrocarbons, forming a pasty mass. These properties would appear to render it a valuable ingredient for an explosive, and Salvati quotes Dr. Bertoni as highly recommending its use. Notwithstanding its high boiling point, however, tetranitro-methane is extremely volatile, and this very serious defect appears to have been overlooked.

657. III$_2$. **Nitro-molasses** is prepared by nitrating 380 parts molasses with 1,000 parts fuming nitric acid and 2,000 parts concentrated sulphuric acid. The product when washed is a grey yellow or whitish precipitate. By a special treatment a liquid nitro-product is obtained in lieu of a solid one. (Spec. No. 1,883, 13.4.83.)

658. III$_2$. **Nitro-naphthaline** is formed by the action of nitric acid on naphthaline ($C_{10}H_8$); mono-, di-, tri-, and tetra-nitro-naphthaline are known. *See* p. xl.

659. **Nitro-petrol.** *See* NITRO-XYLOL.

660. III$_2$. **Nitro-phenol** is produced by the action of nitric acid on phenol (phenyl alcohol, carbolic acid, &c.) C_6H_6O. Mono-, di-, and tri-nitro-phenol are known. The latter is termed picric acid. *See* pp. xxxvi and xl.

661. **Nitro-pyline.** *See* VOLKMAN.

662. III$_2$. **Nitro-peat** is the result of the action of the usual acids on peat. The humus of the peat is converted into a dark brown sticky liquid like that obtained by nitrating the heaviest tar oils, while the finely-divided vegetable fibres are converted also into a nitro-com-

Nitro-peat—*continued.*

pound. Peat of recent formation give a violent action with the strong acid, and a preferable description is found in a firm solid peat of somewhat ancient formation. Like nitro-tars and similar products, nitro-peat has about the same specific gravity as water. It has a powerful aromatic odour, especially when burnt. It burns in the open air with a smoky flame. (M. No. II., p. 453.)

662*. **Nitro-resorcin.** *See* HAUFF.

663. III_2. **Nitro-saccharose** is produced by nitrating sugar. It is a white, sandy, explosive substance, soluble in alcohol and ether. That made from cane sugar does not crystallise from solution, while that made from milk sugar does. It has been used in percussion caps, being even stronger and quicker than nitro-glycerine, but owing to difficulties inherent in its manufacture, its great sensitiveness, and its hygroscopic qualities, combined with its proneness to decomposition, it is not used, by itself at least, as an explosive for blasting. *See also* GLUKODINE. (D., p. 671, and M. No. II., p. 446.)

664. III_2. **Nitro-starch,** called also Xyloïdine and Pyroxylam, is a white powder produced by the action of the usual acids on starch. The starch cannot be put directly into the mixture of acids, because it would form clots, which would not be entirely acted upon. Uchatius's process is to dissolve one part of potato starch in eight parts of fuming nitric acid, keeping it cool. The resulting syrupy solution is poured in a stream into 16 parts of concentrated sulphuric acid and kept stirred. The mixture is left to stand 12 hours, then washed, treated with a boiling solution of carbonate of soda, and dried.

This powder is very hygroscopic, insoluble in water and alcohol, but soluble in ether. It readily decomposes spontaneously. When dry it is very explosive and takes fire at about 350° F. It does not appear to have come into practical use. (D., p. 670.)

665. **Nitro-sugar.** *See* NITRO-SACCHAROSE.

666. III_2. **Nitro-tar** has been made from crude tar oils by direct nitration with strong nitric acid. The nitro-substances were washed, dried, and mixed with alkaline nitrates, chlorate of potash, or similar substances. The ordinary coal-tar was also tried, but it became evident that its treatment with strong nitric acids was dangerous, and that the process on a large scale was likely to fail. But satisfactory nitro-compounds were obtained by the use of a weaker acid (S.G. 1·53 to 1·45) which when mixed with nitrates, &c., gave a good explosive, as did a solution of them in concentrated nitric acid.

Somewhat similar results were obtained from nitro-compounds formed from pitch and liquid hydrocarbons, but they were found to require much more addition of oxygenated bodies than the nitro-compounds obtained from tar or tar oils. (M. No. II., p. 451 *et seq.*) *See also* SCHULTZE and EMILITE.

667. III_2. **Nitro-toluol** or **Nitro-toluene.** Toluol or Toluene C_7H_8 is a coal-tar product. Treated with nitric acid, mono-, di-, and tri-nitro-toluol are produced. *See* p. xl.

668. $III_{1 \& 2}$. **Nitro-xylol** or **Nitro-xylene** is produced by the action of nitric acid on xylol (C_8H_{10}), a coal-tar product identical with petrol. Mono-, di-, and tri-nitro-xylol have been produced. *See* p. xl.

669. $III_{1 \& 2}$, $IV_{1 \& 2}$. **Nobel** has proposed to use a mixture of metallic salts rich in oxygen, *e.g.*, nitrate, chlorate, or perchlorate, with one of the nitro-compounds of glycerine sugar, or cellulose. The barium, potassium, and sodium salts are mentioned, and for blasting operations a mixture of from 75 to 80 per cent. of one of these with 25 to 20 per cent. of nitro-glycerine is recommended. For firearms he proposes the addition of 5 to 15 per cent. of nitro-glycerine, or 10 to 30 per cent. of nitro-glycerine thickened with nitro-cellulose or nitro-sugar, or cellulose alone. (M. No. XIII., p. 246.)

670. V_1. **Nobel** proposes to substitute for fulminate of mercury various explosive mixtures in a finely granulated condition, *e.g.*, 2 parts collodion cotton dissolved in 12 parts acetone, 1 part nitro-glycerine, 4 parts picrate of potash, and 8 parts chlorate of potash. On evaporating the acetone the product is easily broken into grains, 3,000 to 15,000 to the gramme. (Spec. No. 16,919, 12.10.88.)

671. III_1. **Nobel** has proposed a mixture of 100 parts of gunpowder and 40 parts of nitro-glycerine, to be prepared immediately before use. The mixture is enclosed in boxes of sheet zinc, and is intended to ensure the explosion of nitro-glycerine cartridges. (D. 694 and 720.)

672. III_1, IV_1. **Nobel** has also proposed a mixture of nitro-cellulose with nitro-starch or nitro-dextrin, manufactured with the aid of acetone or similar solvent. Picrates, chlorates, and nitrates may also be added. (Spec. No. 6,560, 2.5.88.)

673. III_1. **Nobel** has also proposed to mix nitro-glycerine or nitrates of ethyl or methyl with gunpowder, gun-cotton, or other analogous substances. (Spec. No. 2,359, 24.2.63.)

674. II., $III_{1 \& 2}$. **Nobel** proposed a gunpowder for propelling purposes, composed of 80 parts of saltpetre, six parts, or less, of sulphur, and 14 parts of charcoal. To this was to be added, when desirable, to quicken its action (*a*) ordinary gunpowder, or (*b*) compressed gun-cotton, or (*c*) equal parts of nitrate of potash and picrate of ammonia, with the addition of a little gum. For blasting purposes the foundation was ordinary blasting powder, or that made with nitrate of soda, while the quickening agents were (*a*) equal parts of picrate of lead or potash, and saltpetre, with a little gum, or (*b*) gelatinised nitro-glycerine, or other known nitro-glycerine compound, or (*c*) compressed gun-cotton. He expressly bars the use of chlorate of potash powders as being too dangerous. (Spec. No. 226, 20.1.79.)

675. II. **Nobel** has patented the employment as explosives of combustible liquids holding in solution various nitrates or other salts which readily give up oxygen, with or without the addition of nitrates or analogous salts in fine powder. The resulting product can be mixed with other explosive substances. (Fr. Spec. No. 161,260, 29.3.84.)

676. III_1., IV_1. **Nobel** has also patented:—

(1) The lowering of the freezing point of nitro-glycerine by dissolving other substances in it.

(2) The incorporation with nitro-glycerine of nitrates or chlorates, serving to complete the explosive combination, and to absorb the liquid so as to form a solid explosive.

(3) The incorporation with nitro-glycerine of any other porous matter which is capable of absorbing it. (Fr. Spec. No. 170,290, 24.7.85.)

677. II. **Nobel** has also patented the application as an explosive of nitrate of ammonia without the addition of combustible matter capable of increasing the explosive force, or with the addition of a substance not explosive in itself and not increasing the explosive force.

The nitrate can be fused and cast into cylinders or masses of other form. (Fr. Spec. No. 170,291, 24.7.85, 27.7.85.)

678. II., IV_2., **Nobel** has also patented the employment as propulsive agents of nitrates, nitrites, chlorates, and perchlorates of metallic bases or of ammonia, without the addition of combustible matter. The salts may be either fused and cast, or compressed and granulated. (Fr. Spec. No. 170,292, 24.7.85; No. 170,340, 27.7.85.)

679. VI_2. **Nobel** has also patented the employment in shells and torpedoes of a gaseous explosive in a state of compression, and also of oxygen gas together with a gaseous liquid or solid combustible material. (Fr. Spec. No. 179,289, 27.10.86, 14.6.87.)

680. III_2. **Nobel** adds 3 parts of nitrate to 1 of picrate or ammonia, and hardens with ½ per cent. of gum or dextrine. (Spec. No. 10,722, 24.7.88.) *See* PICRIC POWDER.

He also mixes charcoal and barium nitrate with picrate of ammonia or amorphous phosphorus. (Spec. No. 1,469, 31.1.89.)

681. III_2. **Nobel** has patented an explosive for propulsive purposes, consisting of nitro-mannite incorporated with nitro-cellulose. He claims that this explosive would have the advantage over Ballistite, that there is no liquid ingredient to exude. (Spec. No. 11,645, 2.6.94.)

682. III_2. **Nobel** has also patented powders obtained by dissolving the nitro-derivatives of various celluloses such as corozzo (vegetable ivory), cocoanut fibre, or hard wood, or nitro-starch, or nitro-dextrin. The solvent is evaporated, and the resulting product is then granulated. (Fr. Spec. No. 186,801, 5.11.87.)

683. **Nobel.** *See* BALLISTITE, SAFETY FUZE, DYNAMITE, AMMONIO-NITRATE OF COPPER, &c.

684. IV_2, V_1. **Nobles Powders** were submitted to the French Government Commission in 1880. Of these there were eight samples of the following compositions:—

—		2.	3.	4.	5.	6.	7.	8.
	Parts.	Parts.	Parts.	Parts.	Parts.	Parts.	Parts.	Parts.
Chlorate of potash	100	100	100	50	100	100	100	100
Sugar - -	20	22	24	20	20	20	10	15
Prussiate of potash	16	18	22	8	—	8	16	16
Starch - -	1	—	—	2	2	2	2	1
Camphor -	2	—	—	2	2	2	2	1
Sulphur - -	—	—	—	8	—	—	—	—
Benzoin - -	—	—	—	—	16	8	—	—
Saltpetre - -	—	—	—	—	—	—	50	50
Charcoal- -	—	—	—	—	—	—	10	5

As might be expected they were dangerously sensitive. The addition of the camphor did not have the effect in diminishing the sensitiveness of the compounds containing it which it has in the case of gun-cotton and nitroglycerine compounds.

685. II. **Nordenfelt,** in conjunction with Meurling, proposed to manufacture gunpowder, in which he employed a brittle substance, prepared from cotton or other woody fibre, by treating it with hydrochloric acid gas. This substance is thoroughly mixed with the proper amount of a saturated solution of sulphur in carbon bisulphide, in a closed vessel provided with a mechanical agitator. The dried mixture of carbon and sulphur is saturated with an aqueous solution of saltpetre, and the compound dried and finished as usual.

The brittle substance mentioned above would appear to be hydrocellulose as described by Girard, or fulminose, as described by Blondeau. It is an isomeric modification of cellulose, and is produced by the action of acids, or heat, on that substance. It is extremely brittle. (M., No. I., p. 311, No. X., p. 195.) (Spec. Nos. 6,514–15, 18.4.84, and Watts' Dic. of Chem.)

686. A. III_2. **Normal Smokeless Powder,** adopted by the Swiss Government, is in the form of square greyish leaflets. It is composed of nitro-cotton, gelatinised by a suitable process. A small per-centage of carbonate of calcium may be added.

687. **Norrbin.** *See* AMMONIA DYNAMITE.

688. **Nysebastine.** *See* SEBASTINE.

689. A. III_1. **Oarite,** as authorised for manufacture, is composed as follows :—

Nitrate of potash or baryta	48 parts.
Nitro-cellulose	8 ,,
Di-nitro-benzol	8 ,,
Nitro-glycerine	2 ,,

It was proposed by Trench.

690. **Ohlson.** *See* AMMONIA DYNAMITE.

691. II. & IV_2. **Oliver** proposed to use peat in lieu of charcoal in gunpowder, and also chlorate of potash, mixed with beeswax or tallow, and resin in lieu of sulphur. In this case the ingredients were to be mixed in a damp state. (Spec. No. 1,800, 10.6.69.)

692. **O.P.C. Safety Compound.** *See* ORIENTAL POWDER.

693. III_2. **Orange Lightning Powder,** a sporting powder analogous to E.C. (P. & S. 700.)

694. III_2. **Oriasite** is the same as Meganite, *q.v.*, except that nitrated wood stuff alone is used, the use of vegetable ivory being abandoned.

695. **A.** VII_2. **Oriental Fireworks*** are licensed for importation as consisting of saltpetre, sulphur, and charcoal, enclosed in a paper or bamboo case, with or without the addition of a small charge consisting of not more than two grains of a mixture of realgar and chlorate of potash at one end of the case.

696. II., IV_2. **Oriental Powder** consists of a mixture of tan, bark, sawdust, or other vegetable fibre, impregnated with a nitrate or chlorate, and gunpowder or similar explosive compound. Spent tan is preferred. (T., p. 100.)

697. **Orioli,** in 1881, submitted a so-called smokeless powder to the French Committee on Explosives. His assertions did not appear to be justified. (P. & S. 704.)

698. **Oxalates.** Certain of the metallic oxalates possess explosive properties in virtue of the exothermic reactions to which they may give rise. The oxalates of silver and of mercury detonate when rapidly heated or submitted to violent shock. (B. II. 123.)

699. II. **Oxland's Powder** consists of:—

Nitrate of soda	85	parts.
Sulphur	16	,,
Charcoal	18	,,
Coal or dust (*sic*)	20	,,

The nitrate of soda is refined from the crude salt by precipitating the salts of lime and magnesia by carbonate of soda, and evaporating the filtered solution. (Spec. No. 1,740, 18.7.60., D. 609.)

700. III_1. **Oxonite,** invented by Punshon, is a mixture of picric and nitric acids. The picric acid, sometimes with

* These will in future be termed simply manufactured fireworks, and not separately defined.

Oxonite—*continued.*

the addition of a nitrate, is packed in a calico cartridge, which also contains the nitric acid in a hermetically sealed glass tube. This tube is to be broken by a blow before putting the cartridge into the bore-hole. This explosive caused a serious accident to the patentee, and another in August 1884. They were experimenting in (fortunately for them) bore-holes in soft mud. While tamping a cartridge it exploded, very seriously injuring them. There is no doubt that the accident was due to absence of precaution against allowing the nitric acid to gain access to the contents of the detonator, intended, in conjunction with a fuze, to fire the cartridge. (Spec. No. 2,428, 12.5.83.) *See* PUNSHON (773).

701. **Oxydine.** *See* TURPIN.

702. I. **P. Powder** (Pebble) was introduced into the service in 1871. The grains are approximately cubical in form, and average 80 to the lb. Density 1·75.

703. I. **P² Powder,** introduced into the service in 1876, is approximately cubical in form with a length of edge of about 1″·5. Density 1·75. This powder has now been superseded by E.X.E.

704. **Paine.** *See* DAVIES.

705. VI_2. **Pain's Instantaneous Pyrotechnic Fuzes** consist of low tension electric fuzes in which the fine wire is embedded in gunpowder.

706. III_2. **Paleina** is pulped straw nitrated in thin sheets, and washed. A solution of saltpetre and dextrine with ground hardwood charcoal is added. When dry it is like small discs of cardboard. It absorbs nitro-glycerine. *See* STRAW DYNAMITE.

707. **Paleine.** *See* STRAW DYNAMITE and LANFREY'S POWDER.

708. **Pancera.** *See* DIORREXINE.

709. III_1. **Panclastite** is a name given to various mixtures proposed by E. Turpin. In some he proposes to mix liquid nitrogen tetroxide, or nitric peroxide (N_2O_4)* with bisulphide of carbon, benzol, petroleum, ether, mineral volatile oils, or other liquid or solid hydrocarbons. He recommends especially a mixture of $2CS_2 + 3N_2O_4$, but the use of two such ingredients, both giving off highly deleterious vapours, is a very serious objection to the employment of such a mixture in a mine or other confined space, and the vapour given off by the liquid N_2O_4 is especially dangerous. (Spec. No. 4,544, 1881, and No. 1,461, 27.3.82.)

710. III_1. **Pantopollite,** a cheap dynamite, made at Opladen. It consists of nitro-glycerine simply mixed with naphthaline or nitro-naphthaline. (P. & S. 716.)

711. **Paper Powder.** *See* MELLAND.

712. **Papier Fulminant.** *See* PYRO-PAPIER.

713. III_2. **Paraffined Gun-cotton.** The French Committee on Explosives have carried out experiments with samples of gun-cotton containing from 5 per cent. to 20 per cent. of paraffin, the per-centage being chosen so as to be sufficient to prevent the mass being exploded by the impact of a bullet, but not sufficient to prevent its detonation with the service detonator. (P. & S. 722.) (Memorial des Poudres et Saltpêtres, I. 483, II. 586 and 605, VII. 73.)

714. IV_2. **Parone's Explosive** consists of two parts of chlorate of potash to one part of bisulphide of carbon. It was tried in a 24 c.m. mortar in Italy, which burst at the first round. It is really a form of Rack-a-rock, *q.v.* (M. No. X.V., p. 574.)

---

* N_2O_4 is a gas or vapour which can at low temperatures be solidified. The crystals melt at $-9°$ C., but will not re-crystallise except at a much lower temperature. A little above $-9°$ C., the liquid is colourless, but becomes gradually orange yellow, and boils at 22° C. The vapours are powerfully irritating and highly dangerous.

715. **Parozzani.** *See* PYRO-COTTON.

716. III $_2$. **Patent Gunpowder.** This was a wood gunpowder (nitro-lignin) formerly manufactured at Glyn Ceiriog, in Wales. The loss of the "Great Queensland" in 1876 was held by the Wreck Commissioners Court to be probably due to the presence of this powder on board, which by reason of its impure condition when shipped ignited spontaneously.

717. IV$_2$. **Pattison** mixes chlorate of potash explosives with greasy vegetable flour or bran to lessen the liability to explode. (Spec. No. 810, 24.2.80.)

718. **Pebble Powder.** *See* P. and P². POWDERS.

719. IV$_2$. **Peiey's Explosive Paper** consists of ordinary blotting paper soaked in the following mixture:—

Chlorate of potash	66 parts.
Yellow prussiate of potash	17 „
Refined salt	35 „
Wood charcoal	17 „
Starch	10 „

the whole being mixed with 10 times its weight of boiling water. The paper is dried, cut into strips, and rolled into cartridges. (R.A.I. Proc., Feb. 1886.)

720. I. **Pellet Powder.** Black powder compressed into cylindrical pellets fitting into a small-arm cartridge.

721. III $_2$. **Pellier** has patented, under the name of "explosive resin," a product composed of cane or beet sugar, nitrated in the usual manner, combined with lime, and then treated with strong or dilute hydrochloric acid. The resulting product is a yellowish powder. (Fr. Spec. No. 180,555, 29.12.86.)

722. V$_1$. **Pellier's Powder** was submitted to the French Government Commission in 1884. It consists of:—

Chlorate of potash	100 parts.
Saltpetre	12½ „
Sulphur	12½ „
Fine sawdust	9 „
Extract of logwood	15 „

Pellier's Powder—*continued.*

It is practically the same as Kellow and Short's powder. The same inventor had previously, in 1882, submitted to the same Commission two similar powders composed as follows:—

—	For Blasting.	For Guns.
	Parts	Parts.
Chlorate of potash	44 to 47	40
Yellow prussiate of potash	38 „ 36	15
Flowers of sulphur	18 „ 17	10
Sugar	—	10
Charcoal	—	5
Saltpetre	—	15

Both were of course extremely sensitive.

723. II. **Penniman** has proposed to protect nitrate of ammonia from moisture by an envelope of petroleum or petroleum products. (Fr. Spec. No. 166,946, 10.2.85.)

724. II. **Peralite** is a large grained powder consisting of:—

Saltpetre	63 parts.
Charcoal	30 „
Sulphide of antimony	6 „

725. **Perchlorate of Potash.** $KClO_4$ has been proposed as a substitute for chlorate of potash $KClO_3$, the former being a more stable compound and containing more oxygen. Perchlorates, however, are exothermal, that is, they give out heat in formation, and are therefore not so efficient as ingredients in explosives as chlorates, which are endothermal and disengage heat on decomposition.

726. **A.** VI$_1$. **Percussion Caps** are small copper or brass receptacles, containing detonating composition. In England a percussion cap is regarded as a detonator if it contains more than half a grain of composition, owing to the liability to explode *en masse* of caps containing more than this charge. When, however, the proportion of fulminate of mercury does not exceed one quarter of the whole composition, as in the case of the Government cap for cordite, this limit of charge is raised to 0·6 grain. All caps must be treated as detonators when heated in the process of drying, or for any other purpose. *See* CAP COMPOSITION.

727. I. **Père** has proposed as a substitute for charcoal flax-straw, carbonised in fixed or revolving cylinders. (Fr. Spec. No. 179,380, 5.11.86.)

728. V_1. **Perkins** proposed the use of amorphous phosphorus with metallic sulphides, especially of antimony, and chlorate or nitrate of potash, for fulminating powders. (Spec. No. 898, 28.3.70.)

729. V_2. **Permanganates.** Some of these are said by Muthman to be extremely explosive by percussion, especially permanganate of ammonia. (A. & E. II. 6.)

730. V_1. **Pertuiset's Powder** consists of :—

Chlorate of potash - - -	2 parts.
Sulphur - - - -	1 ,,
Sporting gunpowder - -	$\frac{1}{8}$,,
Animal charcoal - - -	$\frac{1}{50}$,,

It is recommended for use mainly as a detonating powder, or for filling bullets or shells. (Spec. No. 2,837, 9.10.67, and No. 2,066, 21.7.70.)

731. **Pesci.** *See* MAÏZITE.

732. III_2. **Petragit.** Proposed by Doutrelepont and Schreiber, consists of :—

Nitro-saccharose (molasses) -	38·6 per cent.
Nitro-lignin - - -	5·0 ,,
Nitrate of potash - -	56·4 ,,

It is claimed that this mixture is unfreezable. (H. M.)

733. **Petralit.** *See* LÏESCH.

734. III_2. **Petralite** consists of :—

Nitrate of soda or potash - -	64 per cent.
Nitrated wood or charcoal -	30 ,,
Carbonate of ammonia - -	6 ,,

It was made at Fahlun in 1879. ("Colliery Guardian," 22.6.89.)

735. III_1. **Petralithe** consists of :—

Nitro-glycerine - -	640 parts.	
Nitrate of ammonia, soda, or urea - -	120 ,,	
Palmitinate of cetyl -	$2\frac{1}{2}$,,	(the principal part of the solid portion of spermaceti.)

Petralithe—*continued.*

Carbonate of lime - -	$2\frac{1}{2}$ parts.
Prepared vegetable or animal charcoal.	230 „
Bicarbonate of soda - -	5 „

This somewhat curious mixture was proposed for adoption in England in 1882, and was patented, but no active steps were taken for its introduction. (Spec. No. 2,302, 25.5.81.)

This compound as submitted to the French Government Commission was stated to consist of:—

Nitro-glycerine - - -	60 parts.
Nitrate of potash, soda, or ammonia	16 „
Spermaceti - - - -	1 „
Carbonate of lime - -	1 „
Lignin - - - -	6 „
"Special" charcoal - -	16 „

(Spec. No. 2,149, 25.5.80.)

736. V_1. **Petrofracteur** consists of:—

Nitro-benzol - - -	10 per cent.
Chlorate of potash - - -	67 „
Nitrate of potash - -	20 „
Sulphide of antimony - -	3 „

It generally resembles Kinetite, *q.v.*, but contains no nitro-cotton. It was favourably reported upon by an Austrian Military Committee. (Journal of Soc. Chem. Ind., Vol. VI., p. 5.)

737. **Petry.** *See* DYNAMOGEN and KINETITE.

738. III_2. **Picrates.** *See* INTRODUCTION, p. xxxix.

739. III_2. **Picrate of Potash** ($C_6H_2(NO_2)_3KO$) has been tried in Austria and America as a charge for shells, but by itself it does not contain sufficient oxygen to burn up the whole of its carbon, and hence requires the addition of some oxydising agent. It is a gold yellow substance, crystallising in needles. A violent blow or the contact of flame explodes it. Heated gradually to about 600° F. it detonates violently. When containing 15 per cent. of moisture it is safe from ignition by a blow, and only ignites locally on contact of flame. (D., p. 737.)

740. A. III_2. **Picric Acid.** *See* p. xxxvi.

741. III_1. **Picric Nitro-gelatine** consists of a mixture of nitro-glycerine with 10 per cent. of its weight of picric acid, the whole being gelatinised by means of soluble nitro-cotton. It is intended for use by itself or with another explosive. (German Patent, 1887.)

742. III_2. **Picric Powder** is practically the same as Brugère's. It was proposed by Abel for filling shells, and consists of three parts of saltpetre to two parts of picrate of ammonia. It takes a very violent blow to explode it. Unconfined, and on contact of flame, it only burns locally. It requires strong confinement to develop its force. (D., 740.)

743. III_2. **Pieper** proposes to mix nitro-hydrocarbons with nitrate of ammonia by means of a common solvent. (Spec. No. 23,773, 9.12.93.)

743a. **Pieper.** *See* THORN, WESTENDARP.

744. **Pietrowicz.** *See* SILESITE.

745. **Piquet.** *See* POCHEZ.

746. **Pistol Powder.** *See* RIFLED PISTOL POWDER.

747. A. III_2. **Plastomenite,** a smokeless powder defined as "consisting of thoroughly purified nitro-cotton mixed or "incorporated with thoroughly purified dinitro-toluol "and nitrate of barium." Three varieties are manufactured, viz.:—

J.C.P. - -	- Sporting powder,
B.P. - - -	- Rifle powder,
K.M.P. - -	- Army powder.

748. III_2. **Plera,** a fancy name given to Rifle Gun-cotton.

749. **Plympton.** *See* ACETYLIDES.

750. **Poch.** *See* PUDROLITHE.

751. III_2. **Pochez and Piquet** have patented a process of manufacture of nitro-cellulose from animal refuse. (Fr. Spec. No. 146,181, 3.12.81.)

752. IV_2. **Pohl's Powder** is almost identical with Augendre's. It consists of :—

Chlorate of potash	-	49 per cent.
Yellow prussiate of potash	-	28 ,,
Cane sugar	-	23 ,,

(D. 615.)

753. III_2. **Polis** has proposed an explosive consisting of ditoluol-nitrate of lead, which is obtained in the form of a white amorphous powder. It explodes readily when heated. (P. & S. 769.)

754. II., IV_2. **Pollard** has suggested a method of manufacture which consists in treating sulphur with melted paraffin so that each particle becomes coated with this hydrocarbon, and then incorporating with a chlorate or nitrate. (A. & E. I. 35.)

755. **Polynitro-cellulose.** *See* HEUSSCHEN.

756. III_1. **Porifera Nitrolium** is a mixture of nitro-glycerine with sponge or vegetable fibre, with or without the addition of plaster of Paris.

757. III_1. **Potentia,** an American form of dynamite of the No. 2 type. (T., p. 88.)

758. **A.** III_2. **Potentite** is a nitrated gun-cotton identical with Tonite, except that saltpetre is generally used instead of nitrate of baryta.

It was originally called Liverpool Cotton Powder.

759. **Prado** has patented slagwool, with or without other substances, for the transport and storage of explosive and inflammable material. (Fr. Spec. No. 181,307, 2.2.87.)

760. **Preisenhammer** proposed a mixture of hydrogen and oxygen gases for blasting. (Spec. No. 3,377, 23.9.61.)

Preisenhammer—*continued.*

This idea has been revived by Edison for land mines (M. No. VII., p. 114), and also, I believe, from time to time by other persons. Of course, if solid or liquid mixtures of these gases could be preserved till required for use we should have a most powerful and valuable explosive, but unfortunately the difficulties in the way of such a course appear, to our present knowledge, insuperable.

761. III_2. **Prentice** proposed to regulate the speed of burning of gun-cotton by weaving or interlacing yarns of inert cotton with yarns of gun-cotton, or by pulping the two together. He suggested a paper for sporting purposes made of 15 parts of unconverted to 85 parts of converted fibre. About 30 grains of such paper rolled into a cylinder was suggested as a proper charge for 1 oz. of shot. (Spec. No. 953, 3.4.66.)

762. III_2. **Prieur Powder,** an explosive of which the principal ingredient is a salt called by the inventor trinitro-homophenolate of ammonia. (P. & S. 780.)

763. **A.** III_1. **Primers for Gelatines** have been licensed as consisting of not more than 50 parts of nitro-glycerine mixed with 50 parts of nitro-cellulose, with or without carbonates of calcium and magnesium.

764. I. **Prism[1] Black Powder** was introduced into the service in 1881. It is moulded into regular hexagonal prisms, 0″·9764 in height and 1″·3662 in diameter over the sides. Each prism has a circular hole through its axis. The powder is made from ordinary black grain pressed to a density of 1·76. Prism[2] Powder is obsolete and differed only from Prism[1] in that the prisms were of larger size (2″ + 2″·35).

765. I. **Prism[1] Brown Powder,** introduced in 1884, is of the same dimensions as Prism[1] Black. The proportions of the ingredients are :—

Saltpetre - - - -	79 per cent.
Sulphur - - -	3 „
Charcoal - - - -	18 „

The brown colour is due to the slack-burnt charcoal used. Density 1·8.

766. **Prodhomme.** *See* PYRONITRINE.

767. **Progressite.** *See* TURPIN (980).

767a. **Progressite,** another explosive of this name proposed by the Carbonite Syndicate (Limited), consists of:—

Nitrate of ammonia	- -	94 per cent.
Hydrochloride of aniline	- -	6 „

The ingredients are mixed together with the aid of water. It is stated to be less hygroscopic than other nitrate of ammonia explosives, and to be safe for use in fiery mines. (A. & E. III. 6.) A sample of this explosive is now (December 1894) under examination with a view to its being put on the authorised list.

768. **Promethean Fuze.** *See* COLLIERY SAFETY LIGHTERS.

769. **Prussian Fire.** *See* WIGFALL.

770. **A.** II. **Pudrolithe,** or **Rock Powder,** consists of:—

Saltpetre	- -	68 parts.		
Sulphur	- -	12 „	or { Sulphur - -	8 parts.
			Powdered gumlac	3 „
Charcoal	- -	6 „		
Nitrate of baryta		3 „		
„ soda	-	3 „		
Sawdust	- -	5 „		
Spent tan	-	3 „		

The nitrates of soda and baryta are dissolved in hot water and the tan and sawdust added to the solution and boiled until dry The other ingredients are then added, and the whole mixed together. The compound claims to be slow-burning, and to give but little smoke.

The manufacture of this explosive was carried on at a factory near Llangollen, but was abandoned a few years ago. (Spec. No. 656, 2.3.72.)

771. I. **Pulverin,** a special non-granulated powder made in France for use in mines. Its composition is:—

Saltpetre	- - -	75·0 per cent.
Sulphur	- - - -	12·5 „
Charcoal	- - -	12·5 „

(P. & S. 790.)

772. III_1. **Punshon's Explosive** is a mixture of nitro-glycerine with carbonised or charred peat as an absorbent, in the proportion of 70 parts of nitro-glycerine to 30 of peat. The nitro-glycerine "is cleaned by means of chalk " mixed with water instead of by the use of alkalies " . . . . When as much nitro-glycerine has been " absorbed by the peat as it is capable of holding, finely " ground gun-cotton may be added in such proportion as " will ensure the detonation of the prepared nitro-" glycerine at a temperature at which ordinary nitro-" glycerine would rarely explode." This last paragraph is rather obscure, looking at the respective exploding points of nitro-glycerine and gun-cotton. (Spec. No. 4,268, 9.12.75.)

773. III_1. **Punshon** absorbs nitric acid in asbestos, or other porous substance, and picric acid. A paste is made and put up in paper cartridge cases lined with a cement of ground glass and concentrated solution of silicate of soda to prevent the action of the acid on the paper. (Spec. No. 2,242, 1.6.80.) *See* also OXONITE.

774. III_2. **Punshon's Gun-cotton** was prepared with a view of regulating its rapidity of explosion. The gun-cotton was soaked twelve hours in a solution of sugar preferably refined or crystallised. He also added what he termed "a flux or an explosive, such as white or ordinary " gunpowder, or nitrate of soda or potash, or ordinary " gun-cotton, or a gun-cotton containing from 5 to 20 per " cent. of nitrate of potash or nitrate of soda, combined " with equal or double the same weight of sugar." The object of this was to aid the development of the explosive force. The resulting mass was cut into suitable sizes. This explosive is simply nitrated gun-cotton mixed with sugar. (Spec. No. 2,867, 31.10.70.)

775. **Punshon.** *See* VICTORITE.

776. III_2. **Pyro-cotton** proposed in 1883 by Parozzani for charging shells. It consists of a mixture of gun-cotton with picrates and other substances.

777. III_1. **Pyro-glycerine,** a synonym for nitro-glycerine.

778. II. **Pyrolithe** consists of :—

Saltpetre	51·5	per cent.
Nitrate of soda	16·0	,,
Sulphur	20·0	,,
Sawdust	11·0	,,
Charcoal	1·5	,,

The inventor (Matteen) also patented another mixture, consisting of :—

Nitrate of soda	47	per cent.
Saltpetre	18	,,
Sulphur	17	,,
Sawdust	12	,,
Carbonate or sulphate of soda	6	,,

The idea is to have no carbonic oxide in the products of combustion. (D. 606.)

779. II. **Pyronitrine** was submitted to the French Government Commission in 1884, by M. Prodhomme. There were two samples of the following compositions :—

—	1.	2.
	Per Cent.	Per Cent.
Nitrate of soda	35	18
Saltpetre	35	45
Tan	15	15
"Sulphate"	2	3
Sulphur	6	9
Charcoal	3	—
Resin	4	3
Tar	—	7

(Spec. No. 4,200, 15.10.80.)

780. V_1. **Pyronome,** invented by Sandoy, consists of :—

Saltpetre	69	parts.
Sulphur	9	,,
Charcoal	10	,,
Metallic antimony	8	,,
Chlorate of potash	5	,,
Rye flour	4	,,

together with a small quantity (a few centigrams) of chromate of potash. The ingredients are boiled together, the mass evaporated down to a paste, dried and powdered. (Spec. No. 3,923, 9.9.81.)

781. **Pyronome.** *See* DE TRET.

782. III_2. **Pyropapier** or **Papier Fulminant** is made by immersing for two minutes unsized paper into equal parts of nitric and sulphuric acids. It is washed, treated with an ammoniacal solution, re-washed, and dried. It has been employed as a primer for the needle gun. (D., p. 667.) *See* DYNAMOGEN, HOCHSTÄTTER, MELLAND.

783. **Pyroxylam.** *See* NITRO-STARCH.

784. III_2. **Pyroxiline,** or **Pyroxyle,** or **Pyroxylol,** is a generic name for all the nitro-substances formed from various forms of cellulose. It includes, for instance, gun-cotton, gun-paper, wood gunpowder, &c.

785. III_2. **Pyroxilite,** a powder proposed by Anthoine and Grouselle in 1887. It consists of picric acid, oxide of lead, and bichromate of potash or chromic acid. (P. & S. 806.)

786. I. **Q.F. Powder.** A black powder, introduced into the Service in 1887 for quick-firing guns of small calibre. It is analogous to the French C_2 powder. The grains are cubical in shape and number 180 to 200 to the lb. Density 1·75.

787. **Quentin.** *See* COMBUSTIBLE CORD.

788. I. **Quick's Powder.** A powder moulded in the form of discs on perforated cakes, with the object of ensuring a constant surface of combustion of the charge. (P. & S. 810.)

789. **Qurin.** *See* SCHULHOF and QURIN.

790. **R.** III_2. **Rack-a-rock** consists of cartridges made of compressed chlorate of potash, which are dipped when required for use into certain liquids, which are by themselves, like the chlorate of potash, of an inexplosive character. The liquids are "dead oil," a dark heavy oil consisting chiefly of hydrocarbons derived from coal tar and having a high boiling point, or a mixture of dead oil with its own volumes of bisulphide of carbon, or a mixture like the last with the addition of about 3 per cent. of sulphur.

Rack-a-rock—*continued.*

The chlorate of potash cartridges are enclosed in small bags of cotton, or other cloth of suitable size and shape, and dipped into the liquid. Preferably the mixture of dead oil and bisulphide of carbon is used, the object of the bisulphide of carbon being to prevent too large a proportion of dead oil from being taken up by the chlorate. The bisulphide afterwards evaporates out.

Nitro-benzol is also used as a dipping liquid, or as an ingredient thereof in combination with picric acid.

In practice the cartridges in a wire basket are suspended from a spring balance and dipped into a pail containing the fluid. The completion of the absorption of the proper amount of fluid is shown by the spring balance. The proportions given are 3 to $4\frac{1}{6}$ parts solid to 1 part liquid ingredient.

In the great Hell Gate explosion, which took place on the 10th October 1885, 240,399 lbs. of rack-a-rock were used in conjunction with 42,331 lbs. of dynamite.

If the cartridges be kept they appear to tend to increased sensibility to friction or percussion. (Spec, No. 5,584, 21.12.81; No. 5,596, 21.12.81; No. 1,461, 27.3.82; Nos. 5,624–5, 4.12.83.)

791. **A.** VI_1. **Railway Fog Signals** are usually flattish circular tin or tinned iron cases, containing a small quantity of powder, and nipples armed with ordinary percussion caps. They are secured to the rails by soft lead clips, and the weight of a locomotive wheel passing over them fires the caps, and consequently the powder. They are in this country required to be of such strength and construction that the explosion of one will not cause the explosion of similar ones contiguous to it.

792. IV_1. **Rand** has proposed the following mixtures :—

—	A.	B.	C.
	Per Cent.	Per Cent.	Per Cent.
Nitro-benzol - - -	20	15	15
Chlorate of potash - - -	80	42·5	51
Manganese binoxide - -	—	42·5	—
Some inert substance - -	—	—	34

(Spec. No. 12,744, 12.7.92.)

793. **Randite.** *See* RAND.

794. IV_2. **Rave** proposed to press a current of chlorine produced by the action of hydrochloric acid on di-oxide of manganese into a mixture of 80 parts of carbonate of potash, 30 parts of ground straw, and 15 parts of anthracite, made into a thin paste with water. The resulting mixture was to be dried and used as a blasting powder. Of course this is simply a chlorate of potash mixture obtained in a roundabout way. (Spec. No. 2,469 23.11.59.)

795. IV_2. **Raves' Powder** is a mixture of chlorate of potash and charcoal, wood dust, or coal. The ingredients are made into a paste, dried and ground. The average proportions are given as 100 parts of vegetable matter or carbon to 200 parts of chlorate of potash. (Spec. No. 2,651, 23.11.59.)

796. III_2. **Reeves** proposed to obtain a graduated series of explosives by immersing his material (cellulose in various forms) in a bath of one part of nitric to two parts by volume of sulphuric acid. The first lot was left in the bath 22 to 26 hours; the second lot 32 to 60 hours, about one-twentieth of the original amount of nitric acid being added; the third lot 2½ to 4 days, 15 to 25 per cent. of the original amount of nitric acid being added. He stated a fourth and fifth lot might be dipped in the bath either with or without the further addition of nitric acid.

The nitro-cottons thus obtained were to be mixed as required. (Spec. No. 989, 2.4.67.) *See also* MUSCHAMP.

797. IV_2. **Reichen** patented some rolls or cartridges of paper impregnated with a chlorate mixture identical with Melland's Paper Powder. (Spec. No. 2,266, 2.9.65.)

798. III_2. **Reid** and **Johnson** have patented the hardening of the grains of powders containing nitro-cellulose or other solid nitrated organic bodies. The powder proposed consists of 100 parts by volume of nitro-cellulose, moistened with 50 to 80 parts of ether, ethylic, or methylic alcohol, or a mixture of these with each other or with other liquids. (Fr. Spec. 147,325, 11.2.82.)

799. III $_1$. **Rendrock** is a mixture of an alkaline nitrate with nitro-glycerine, wood fibre, and paraffin, or similar substance. The suggested proportions are :—

Nitrate of potash - - -	40 per cent.
Nitro-glycerine - - - - -	40 „
Wood fibre - - -	13 „
Paraffin (or pitch) - - -	7 „

with or without the addition of small quantities of sulphur and charcoal. (T., p. 101.)

800. II. **Reuland** has patented explosive mixtures consisting of naphthaline melted with—

(*a.*) Nitrate of ammonia (preferably made with the aid of sulphate of ammonia) and nitrate of strontia.

(*b.*) Humate or ulmate of ammonia obtained from peat or similar bodies by washing, supersaturating with an alkaline carbonate, and neutralising with an acid.

These explosives are to be used alone or with other hydrocarbons, oxydising bodies, or substances charged with hydrocarbons. (Fr. Spec. No. 215,261, 1.8.91.)

801. II. **Reunert** has proposed the employment of a revolving drum containing copper balls, in which the following ingredients of saltpetre, sulphur, charcoal, and wheat-flour, or starch, are incorporated together under steam pressure. The paste so obtained is then dried and granulated. *See* STARCH POWDER. (Fr. Spec. No. 161,776, 28.4.84.)

802. **Reunert.** *See* FITCH.

803. IV $_2$. **Reveley's Powder** consists of :—

Chlorate of potash - - -	48 per cent.
Yellow prussiate of potash -	29 „
Loaf sugar - - - - - -	23 „

It is simply " White Gunpowder," the proportions of which, however, vary somewhat, those given in Bloxam's "Chemistry" being respectively two parts, one part, and one part of the above-mentioned ingredients.

Augendre's and Pohl's are the same mixture, Augendre's answering to that given by Bloxam, Pohl's being identical with Reveley's. (D. 617.)

804. IV_2. **Reynold's Powder** consists of 75 per cent. of chlorate of potash with 25 per cent. of "sulphurea." It is a white powder, igniting at a lower temperature than gunpowder, and is easily made by the mixture of the two ingredients. "Sulphurea" (sulpho-carbamide $C.S.N_2H_4$), it is said, can be prepared in large quantities from one of the waste products in gas manufacture. (T. 62.)

805. I. **R.F.G., R.F.G.$_2$ Powders.** Two grades of rifle fine grain powders in the Service. The size of grain is 12 to 20 mesh. The powders differ from each other only in density, that of the former being 1·6 and of the latter 1·72 to 1·75.

806. III_1. **Rhenish Dynamite Co.** This company, through Robert Gottheil, proposed in 1874 to employ in place of pure nitro-glycerine a compound consisting of a solution of a hydrocarbon in nitro-glycerine. Naphthaline is recommended as suitable. The solution is prepared by dissolving 2 to 3 per cent. of naphthaline in nitro-glycerine under the action of heat in a water bath, or by mixing a small quantity of kieselguhr saturated with melted naphthaline with the nitro-glycerine and other substances forming the explosive compound.

Two mixtures are suggested, viz :—

Washed infusorial earth (*i.e.*, kieselguhr)	23 or	20	per cent.
Chalk - - - - - -	2 ,,	3	,,
Heavy spar - - - - - - -	— ,,	7	,,
Solution of naphthaline in nitro-glycerine	75 ,,	20	,,

The advantages claimed are somewhat curiously expressed :—"The nitric acid (*sic*) contained in the nitro-"glycerine is completely utilised, as a chemical com-"bination of the hydrocarbon with the nitro-glycerine "is thereby effected instead of a mere mechanical mixture "of the nitro-glycerine with an organic substance, as in "compounds heretofore employed. . . . . (Spec. No. 1,566, 4.5.74.)

807. III_1. **Rhexite** is a mixture of nitro-glycerine, wood fibre, and saltpetre, somewhat similar to Atlas Powder. It is made by the Borkenstein Company in Styria.

Rhexite—*continued.*

A form of this manufactured at St. Lambrecht, in Styria, consists of—

Nitro-glycerine	67	per cent.
Wood meal	4	,,
Decayed wood	11	,,
Nitrate of soda	18	,,

(O. G.) *See also* COAD'S EXPLOSIVE.

808. IV_2. **Ricker's Powders** consist of ten varieties of chlorate mixture containing a number of substances in addition to the chlorate of potash, such as charcoal, half calcined sea-grass, powdered coal, sawdust, nitrates of soda, lead, or potash, wheat flour, bicarbonate of soda, powdered bark, and dried coffee grounds.

The ingredients are all boiled together in water. (Spec. No. 3,297, 9.12.62.)

809. I. **Rifled Pistol Powder,** in use in the Service, is made from the siftings of R.F.G. Powder, the size of grain being 20 to 30 mesh.

810. A. III_2. **Rifle Gun-cotton** is gun-cotton mixed or not mixed with a nitrate, other than nitrate of lead, and mixed with any one or more of the following substances, viz., pure beeswax, paraffin, shellac, gums or resins dissolved in ether, alcohol, and benzoline, such substances to be free from mineral acid.

811. A. III_2. **Rifleite** has been authorised under the following definition:—"Thoroughly purified nitro-lignin, dis-"solved in a safe and suitable solvent, with or without "nitro-benzol and dinitro-benzol, mixed or impregnated "with a nitrate or nitrates (other than nitrate of lead and "ammonium nitrate), or not so mixed or impregnated, and "with or without the addition of graphite." It is in the form of thin brown tablets. Analysis of two samples gave:—

—	I.	II.
	Per cent.	Per cent.
Nitro-lignin	93·68	86·74
Dinitro-benzol	—	10·96
Nitrate of potassium	2·32	2·30
Matters soluble in ether	0·85	—
Volatile matter	3·15	—

See SMOKELESS POWDER (915).

812. **Ripplene,** invented by G. A. Brevester in Australia, is probably a Sprengel explosive.

813. I. **R.L.G., R.L.G.$_2$, and R.L.G.$_4$ Powders,** are large grain powders in use in the Service. They differ from each other principally in size of grain:—

—	Size.	Density.
	Mesh.	
R.L.G. - - - - -	4 to 8	1·7
R.L.G.$_2$ - - - -	3 to 6	1·65
R.L.G.$_4$ - - - - -	1 to 2	1·65

814. II. **Robandis Powder** (or Brise-rocs) consists of:—

Nitrate of potash - - - -	70 parts.
Nitrate of soda - - -	20 „
Sulphur - - - - - - -	15 „
Salt - - - - -	1 „
Coal - - - - - - - -	5 „
Spent tan and sawdust - -	15 „

(O. G.)

815. IV$_2$. **Roberts' Powder** is White Gunpowder, made by a wet process, and kept in the form of a thin paste, to which a per-centage of glycerine is added to prevent it from being dried. The mixture is exploded by detonating an initial charge of the powder (or other high explosive) in a dry state in contact with the pasty mass which forms the main charge. *See* AUGENDRE. (Spec. No. 926, 14.3.73.)

816. II. **Roberts'** and **Dale's Powder.** In this powder nitrate of soda is used wholly or partially in place of saltpetre in gunpowder. A proportion of anhydrous sulphate of soda (prepared by heating sulphate of soda), or of anhydrous sulphate of magnesia, not exceeding 18 per cent. of the nitrate of soda, is used therewith. (Spec. No. 139, 18.1.62.) (D. 605.)

817. IV$_2$. **Robertson** treated nitro-cellulose with a solution of chlorate of potash and coated the result with collodion. He proposed to use this in the form of a covering for a gunpowder cartridge (Spec. No. 2,601, 18.10.61.)

818. A. III$_2$. **Roburite** is an explosive due to Dr. C. Roth, and is now being manufactured on a commercial scale in this country. It has been for some time in use in Germany. It essentially consists of a mixture of nitrate of ammonium with chlorinated dinitro-benzol, and thus much resembles Bellite and Securite. It is a brownish yellow powder, with the characteristic smell of nitro-benzol.

In the license it is defined as consisting of a mixture of:—

(*a.*) Nitrate of ammonia with or without an admixture of nitrate of sodium and neutral sulphate of ammonium, or either of them, provided that the amount of nitrate of sodium so added shall in no case exceed 50 per cent. of the total amount of nitrates present; and

(*b.*) Thoroughly purified chlorinated dinitro-benzol with or without the addition of thoroughly purified chloro-nitro-naphthaline and chloro-nitro-benzol, provided that such chlorinated dinitro-benzol shall not contain more than four parts by weight of chlorine to every 100 parts by weight of chlorinated dinitro-benzol, and that the proportions of chloro-nitro-naphthaline and chloro-nitro-benzol shall not amount to more than 2 per cent. and 5 per cent. respectively of the finished explosive.

This definition succeeded a much simpler one, in which the two main ingredients alone were designated.

The meta-dinitro-benzol $C_6H_4.(NO_2)_2$ is obtained by the action of nitric acid on benzol (C_6H_6) or on nitro-benzol ($C_6H_5NO_2$), and the chloro-nitro form is represented by $C_6H_3Cl(NO_2)_2$. Roburite volatilises without explosion or ignition when slowly heated, and burns slowly in the open, at all events in small quantities. It requires a powerful detonator to develop its force.

Roth gives the following formula for the explosion of Roburite:—

$$C_{12}H_3Cl(NO_2)_2 + 9\,(NH_4\,NO_3) = 6\,CO_2 + 19\,HO + 20\,N + HCl.$$

In order to render Roburite and similar explosives more susceptible to detonation, Trench proposes to mix them with nitro-cellulose. (Spec. No. 18,241, 13.12.88.)

819. A. III_2. **Roburite No. 2** consists of Roburite as above defined, with the addition of chloride of ammonium and sulphate of magnesium, or either of them.

820. III_1. **Roca** has patented under the name of "Dynamites with hydro-carburetted nitro-glycerine base" or "Lithoclastites," explosives consisting essentially of nitroglycerine mixed with combustible substances capable of giving up hydrogen and carbon, but not in themselves explosive. (Fr. Spec. No. 165,487, 20.11.84.)

821. VI_1. **Roca** has also patented a safety fuze under the name of "Meches sans Poudre," of which the core consists of vegetable fibre steeped in nitrates or chlorates. (Fr. Spec. No. 181,019, 20.1.87.)

822. **Rock Powder.** *See* PUDROLITHE.

823. IV_2. **Rogers** proposed a mixture of:—

Chlorate of potash	- - - -	5 parts.
Cascarilla bark -	- - -	2 ,,
Corundum -	- - - - -	3 ,,
India-rubber solution	- - -	3 ,,

for making blasting fuzes. The object of the corundum or other non-inflammable ingredient is to retard and regulate the burning. The compound was to be dipped in bisulphide of carbon, benzol, or other solvent to mix it. The solvent was evaporated, and the mass was ready for making coils or strands. (Spec. No. 1,356, 12.5.70.)

824. **Rollason.** *See* BARNWELL.

825. R. IV_2. **Romit** is a Swedish explosive consisting of a mixture of nitrate of ammonia and naphthaline (or nitronaphthaline) with chlorate and nitrate of potash. It was submitted for licensing in 1888, but rejected on account of its chemical instability, due to action between the nitrate and chlorate. Some of this explosive ignited spontaneously between Sweden and Germany while being conveyed by the inventor himself for experimental purposes. A sample submitted was found to contain:—

Naphthaline and paraffin	-	12·26 per cent.
Ammonium nitrate	- -	48·80 ,,
Potassium chlorate	- -	38·30 ,,
Moisture	- - -	0·64 ,,

825*. III_2. **Ronsalite.** *See* PIEPER (743). Application has been made to have this explosive placed on the authorised list.

826. V_2. **Rosenboom** and **Mertz** have proposed a cartridge containing four glass tubes, which are broken by the explosion of the powder, and bring into contact with one another glycerine, nitric and sulphuric acids, iodide and chlorate of potash. These ingredients are intended to produce nitro-glycerine and iodide of nitrogen. (Fr. Spec. No. 171,127, 11.9.85.)

827. **A.** III_2. **Rosslyn Smokeless Powder No. 1** is defined as consisting of nitro-cotton mixed or impregnated with nitrate of barium and paraffin or vaseline. The explosive is in the form of pellets suitable for rifle or sporting cartridges.

828. **R.** III_1. **Rosslyn Smokeless Powder No. 2** was of the same nature as No. 1, but contained nitro-glycerine. It was rejected on account of the uncertainty in the per-centage of this ingredient.

829. III_2. **Roth** has patented the use of picric acid (or nitro-bodies containing at least 60 per cent. of picric acid) in combination with nitrate of ammonia and fatty drying oils. The explosives so made are to be protected by paper made waterproof by impregnating it with a mixture of essence of terebenthine and solid hydrocarbons. (Fr. Spec. No. 173,550, 15.1.86.)

830. III_2. **Roth** has also patented a process which consists in treating coal-tar and coal-tar derivatives with nitrating and chlorinating bodies, either in separate operations by means of nitric acid or mixtures which liberate nitric acid, and chlorine either free or nascent; or the treatment can be carried out in one operation by employing a mixture of nitrating and chlorinating bodies, such as nitric acid and hydrochloric acid, or nitric acid and chloride of sodium. (Fr. Sec. 177,309, 9.7.86.)

In a "certificate of addition" (12.4.87.) he proposes—

(1.) To add sulphur or nitro-compounds to the chloro nitro-compounds, either alone or mixed

Roth—*continued.*

with suitable oxidising agents, such as nitrate of ammonia.

(2.) To employ benzol, phenol, cresol, or naphthaline, not produced from coal-tar, for the preparation of the chloro-nitro-bodies. *See* ROBURITE.

831. **Roth.** *See* COLLIERY SAFETY LIGHTERS.

832. II. **Rotten** has patented the following:—

Nitrate of ammonia - - -	27 parts.
Naphthaline - - - -	1¼ ,,
Tar oil - - - - -	1½ ,,
Varnish (various) - - -	5 ,,

(Spec. No. 6,258, 3.12.92.)

833. III_2. **Rotten** has also patented:—

Picrate of potash - - - -	25 parts.
Anthracene - - - -	25 ,,
Tar oil - - - - -	4·5 ,,
Varnish (various) - - -	10 ,,

(Spec. No. 6,258, 3.12.92.)

834. **Ruckterschell.** *See* SILOTVAR.

835. III_1. **Rutenberg's Explosive** is simply dynamite made with randanite* instead of kieselguhr. (Spec. No. 360, 30.1.73.)

836. III_1. **Ryves** has proposed the following smokeless powders:—

—	A.	B.	C.
	Parts.	Parts.	Parts.
Trinitro-cellulose - -	50	50	75
Nitro-glycerine - - -	48	48	24
Castor oil - - -	2	2	1
Magnesium carbonate - -	2	2	1
Cotton paper pulp - -	5	8	5

(A. & E., I. 35.)

837. S_1 **Dynamite.** *See* E.C. DYNAMITE.

838. III_2. **S. Powder,** a smokeless sporting powder manufactured in France. It consists of gun-cotton and nitrates of potash and baryta. (P. & S. 804.)

---

* *See* DYNAMITE DE VONGES.

839. A. II. **Safety Blasting Powder** (called also Carbo-azotine) is licensed for manufacture in this country by Messrs. Pigou, Wilks, and Laurence (Limited). In the license it is defined as a mechanical mixture of saltpetre, sulphur, lampblack, sawdust, and sulphate of iron. In the specification the mixtures claimed are:—

Nitrate of potash, ,, soda, ,, lime	one, two, or three nitrates, in all	50 to 64 parts.
Sulphur - - - - - -		13 to 16 ,,
Tanner's bark (that containing refuse animal matter is preferred) or - ; Sawdust, or - - - - - ; Bark and sawdust - - - -		14 to 16 ,,
Soot or lampblack or both - -		9 to 18 ,,

and 5 to 6 parts of sulphate of iron to every 100 parts of the above mixture.

The materials are ground and boiled in a weak solution of sulphate of iron. The compound becomes liquid and then generally solidifies. When nearly solid it is dried.

This powder is issued occasionally in bulk, but more usually in the form of compressed cartridges like those made of gunpowder. It requires compression and confinement before it will explode.

It is recommended to be used as a wash, with about 10 gallons of water to 2 lbs. of the compound as a remedy for *phylloxera vastatrix*. (Spec. No. 3,934, 14.11.74.)

In 1877 the inventor, Cahuc, took out a fresh patent for an improvement on his original powder as described above. He gave the following proportions as the best for blasting in various materials:—

—	Hard Rock.	Less hard Rock and Coal.	Bituminous Coal and Gypsum.
	Parts.	Parts.	Parts.
Saltpetre - - -	70	64	56
Sulphur - - -	12	13	14
Lampblack - -	5	4	3
Spent tan or Sawdust - - - -	13	19	27
Sulphate of iron - -	2	2	5

He claims that these powders are equally powerful with those previously proposed, and safer to manufacture and use. (Spec. No. 4,732, 12.12.77.)

840. A. VI$_1$. **Safety Cartridges** are defined by sec. 108 of the Explosives Act, 1875, to be "cartridges for small-"arms, of which the case can be extracted from the "small-arm after firing, and which are so closed as to "prevent any explosion in one cartridge being com-"municated to other cartridges." For instance, ordinary service small-arm and sporting cartridges are safety cartridges.

841. A. III$_1$. **Safety Dynamite.** Glycerine is mixed with a nitrated hydrocarbon, preferably of the benzol series, 5 to 10 per cent. or upwards of the latter being used. The compound is nitrated in a mixture of 2 parts sulphuric to 1 part of nitric acid, being cooled during the process by the introduction of nitrogen. The resulting product is washed in an alkaline solution at 50° C. and mixed with kieselguhr. Safety in manufacture and against concussion is claimed, and also non-liability to freeze. (Prov. Spec., 3.12.88.)

It has been defined for licensing as consisting of not more than 75 parts by weight of a thoroughly purified mixture of nitro-glycerine and nitro, or dinitro-benzol, or both of them, uniformly mixed with 25 parts by weight . . . . . The remainder of the definition is identical with that for Dynamite No. 1. *See* p. 42 (bottom .

842. A. VI$_1$. **Safety Firing Tubes No. 1** are defined as consisting of a tube of metal or other suitable material containing a percussion cap and suitable mechanical appliances for firing the same.

843. A. VI$_3$. **Safety Firing Tubes No. 2** are similar to the above, with the addition of a priming charge of mealed powder, not exceeding 40 grains. They must not be liable to explosion *en masse.*

844. A. VI$_1$. **Safety Fuze** is defined in Order in Council No. 1, made under Explosives Act, 1875, to be "a fuze "for blasting which burns and does not explode, and "which does not contain its own means of ignition, and "which is of such strength and construction, and con-"tains an explosive in such quantity, that the burning "of such fuze will not communicate laterally with other "like fuzes." It is commonly called "Bickford fuze,"

Safety Fuze—*continued.*

but there are many varieties and several makers, who in most cases have a trade mark consisting of one or more white or coloured threads down through the interior column of powder.

845. VI$_1$. **Safety Fuze.** Nobel proposes to make the core of a thread composed of camphorated blasting gelatine (15 to 20 per cent. camphor), added to a mixture of 70 parts chlorate of potash, 25 of ferricyanide of potash, and 44 of nitro-cotton. When incorporated a soft, india-rubber-like substance is produced, which is easily worked into thread. The advantages claimed are absolute continuity, imperviousness to moisture, and smokelessness. (Spec. No. 1,470, 31.1.88.)

He further proposes, as an improvement on the above, the use of non-volatile solvents such as nitro-naphthaline, nitro-cumol, nitro-benzol, nitro-toluol, nitro-oxytol, acetic ethers of glycerine, etc. He suggests the following composition :—

Nitro-glycerine	100	parts.
Mono-nitro-naphthaline	33	,,
Di-nitro-benzol	33	,,
Di- or tri-nitro-toluol	33	,,
Nitro-cellulose	200	,,
Chromate of potash	9½	,,
Ferricyanide of lead	3	,,

(Spec. No. 20,467, 1893.)

846. **R.** IV$_2$. **Safety Gunpowder** was proposed in 1888 for license in England. It consisted of a mixture of chlorate of potash and glycerine tinged pink. The ingredients, on keeping, separated more or less, and the mixture was rejected on account of its over-sensitiveness after some months' keeping.

847. III$_1$. **Safety Nitro-powder** is an American lignin-dynamite containing nitrate of sodium, and in the lower grades some starch.

848. III$_2$. **Saint Marc Powder**, a smokeless powder in the form of small greenish semi transparent tablets. The exact composition is kept secret, but there is little doubt that it is a gun-cotton powder. It is stated to have been made in imitation of the French Vieille powder.

849. II. **Sala** has patented under the name of "Grenadine," a blasting powder consisting of benzol, glycerine, sulphur, saltpetre, and sand or ashes. (Fr. Spec. No. 147,111, 30.1.82.)

850. II. **Sala** and **Azemar** have patented a blasting powder called Sulphurite, of the following composition:—

Saltpetre - - - -	62 per cent.
Sulphur - - - -	30 "
Charcoal - - - -	4 "
Wood ashes - - - -	4 "

(Fr. Spec. No. 201,319, 14.10.89.)

851. **Salamon.** *See* MOSENTHAL.

852. III_1. **Salite,** an explosive patented in 1878 by Bergenström. It is composed of about 65 per cent. of nitro-glycerine and 35 per cent of nitrate of urea.

853. **Sallé.** *See* PYRONOME.

854. **Saltpetre.** *See* NITRATE OF POTASH, p. xxii.

855. **Sandholite.** *See* LEWIN.

856. **Sandoy.** *See* PYROMONE (780).

857. IV_2. **Saulaville** and **Laligant** make two mixtures, viz.:—

(*a.*) Bisulphate of potash or soda -	36·06 parts.
Nitrate " " " -	28·60 "
Glycerine - - - -	9·20 "
(*b.*) Inflaming matter (calorigène)—	
Chlorate of potash or soda -	50 to 55 "
Coal - - - -	50 to 75 "

The salts are dissolved, the coal admixed, the mixture dried, and the glycerine added. (O. G.)

858. A. III_2. **Sawdust Gunpowder** is a form of nitro-lignin, made, as its name indicates, from sawdust. It is an authorised explosive, but is seldom, if ever, manufactured.

859. A. III_2. **Sawdust** and **Gun-Cotton Powder** consists of a mixture of two or more of the following substances:— Sawdust gunpowder (*see* above), gun-cotton, and cotton gunpowder (nitrated gun-cotton).

860. II. **Saxifragine** consists of :—

Nitrate of baryta - - -	77 per cent.
Charcoal - - - -	21 „
Saltpetre - - - -	2 „

The nitrate of baryta is prepared by treating the chlorate of barium with nitrate of soda. The manufacture is identical with that of gunpowder. To increase the inflammability of the powder the grains, while still damp, are dusted over with meal powder. (D., p. 610.)

861. $III_{1 \& 2}$. **Sayers** mixes nitro- or dinitro-benzol or other nitro-derivatives of hydrocarbons with a suitable proportion of nitrates and gelatinises them with 2 per cent. to 10 per cent. of gun-cotton. (Spec. No. 17,212, 27.11.88.)

862. **Sayers.** *See* LUNDHOLM and SAYERS.

863. I. **S.B.C. Powder** (Slow Burning C.). A brown powder of prismatic form in use in the service for heavy guns. It differs but little from Prism[1] Brown. The density is 1·85. Each prism has a circular indentation in the centre of one face.

864. II. **Schäffer's Powders** consist of :—

Saltpetre - - - -	30 to 38 parts.
Nitrate of soda - - -	40 „
Sulphur - - - -	8 to 12 „
Charcoal - - -	7 to 8 „
Potassio-tartrate of soda (Rochelle salt) $K\ Na\ C_4\ H_4\ O_6$ - -	4 to 6 „

It is a slow-burning powder. (Spec. No. 2,555, 19.10.63.)

A subsequent specification (1866) gives :—

Potassio-tartrate of soda - - -	4 parts.
Saltpetre - - - -	78 „
Sulphur - - - - -	8 „
Charcoal - - - -	10 „

865. I. **Schaghticoke Cubical Powder,** an American black powder resembling S.P. Powder.

866. III$_2$. **Schenker** has proposed for producing a smokeless powder to dissolve nitro-bodies in acetate of ethyl, and to add at the same time a substance soluble in water, and having no chemical action on nitro-bodies (saltpetre, alcohol, &c.). The acetate of ethyl and the neutral body are then extracted with hot water. (Fr. Spec. No. 217,785, 2.12.91.)

866a. IV$_2$. **Schindler** has proposed the following mixtures :—

Chlorate of potash - - -	60 per cent.
Sugar - - - -	15 „
Anthracite coal dust - - -	25 „

(Spec. No. 20,327, 27.10.93.)

867. V$_1$. **Schlesinger's Powder** consists of:—

Chlorate of potash - - - -	3 parts.
Sulphide of antimony - - -	3 „
Flowers of sulphur - - - -	1 „

It was proposed for small arms. (Spec. No. 14,227, 20.7.52.)

868. **Schmidt.** *See* CARBONITE.

869. **Schnebelin.** *See* SCHNEBELITE.

870. R. IV$_2$. **Schnebelite,** a small-arm powder invented by the Abbé Schnebelin. Its composition is as follows :—

Chlorate of potash - -	70·4 per cent.
Cellulose (elder pith or sawdust) -	28·2 „
Starch - - -	1·4 „

The two latter ingredients may be replaced by the flour of farinaceous vegetables, such as rice, potatoes, or beans. (Spec. No. 9,359, 17.2.92. Fr. Spec. No. 217,540, 14.11.91, 25.2.92.)

871. III$_2$. **Schöneweg** has patented :—

(1.) Explosives composed of oxalic acid, or oxalates mixed with gelatinised gun-cotton or other explosive mixtures with the object of increasing the explosive force, of increasing the stability, and of diminishing the flash.

(2.) The employment of an ordinary explosive cartridge surrounded by oxalates, hydro-oxalates, or oxalic acid. (Fr. Spec. No. 183,880, 28.5.87.)

872. **Schöneweg.** *See* SECURITE.

873. **Schreiber.** *See* PETRAGITE.

874. III$_{1 \& 2}$, IV$_{1 \& 2}$. **Schückher** mixes 50 parts of nitro-starch with 40 of nitre and 10 of benzol, or with other nitrates, picrates, or chlorates, or with nitro-benzol. The nitro-starch (xyloidine) is treated with a solvent during the process to harden the finished product. (Spec. No. 11,665, 22.7.89. Fr. Spec. No. 199,734, 22.7.89, 24.3.90.)

875. III$_2$. **Schückher** has also patented a process of making nitro-starch, consisting in dissolving fecula, which has been dried and ground, in nitric acid, and spraying this solution into mixed nitric and sulphuric acids. The precipitate is dried and ground, aniline being added to give stability. (Fr. Spec. No. 208,248, 15.9.90.)

876. **Schückher.** *See* MEGANITE.

877. III$_2$. **Schulhof** and **Qurin** have patented for the manufacture of cannon powder, fuzes, or cartridges, gun-cotton, greased and compressed, coated with collodion or a mixture of collodion and bisulphide of carbon. (Fr. Spec. No. 161,084, 19.3.84.)

878. III$_1$. **Schultze** in 1868 proposed to mix 10 to 60 pounds weight of nitro-glycerine with 100 lbs. of his wood gunpowder. *See* DUALINE. (Spec. No. 2,542, 14.8.68.)

879. **A.** III$_2$. **Schultze Gunpowder** consists of nitro-lignin mixed or impregnated with a nitrate or nitrates (other than nitrate of lead) and with or without starch or collodion (such collodion to consist of nitro-lignin dissolved in ether and alcohol) or solid paraffin free from mineral acid. A sample gave the following proportions :—

Soluble nitro-lignin	24·83	per cent.
Insoluble	23·36	,,
Lignin (unconverted)	13·14	,,
Nitrates of potassium and barium	32·35	,,
Paraffin	3·65	,,
Matters soluble in alcohol	0·11	,,
Moisture	2·56	,,

880. **A.** III$_2$. **Schultze Blasting Powder** consists of the above, with the addition of charcoal.

881. III$_2$. **Schultze** has also proposed the following powders :—

—	Sporting.	Rifle.	Blasting.
	Parts.	Parts.	Parts.
Nitro-tar (or similar nitro-compound).	12	10	15
Pyroxiline - - -	60 to 80	280 to 300	10
Barium nitrate - - -	60 „ 80	100 „ 120	—
Potassium nitrate - -	8 „ 10	40 „ 50	75
Sulphur - - - - -	—	10	10

(M. No. XV., p. 590. D. 667.)

882. III. **Schüpphaus** proposed to add urea to nitro-compounds with a view of increasing their stability. (Spec. No. 22,384, 22.11.93.)

883. III$_2$. **Schwab Powder.** An Austrian smokeless powder composed of pure nitro-cellulose. The grains are coated with graphite. (P. & S. 899.)

884. **Schwarr.** *See* XANTHINE.

885. III$_1$. **Schwartz Dynamite** contains 30 per cent. to 42 per cent. of plaster and sawdust. (P. & S. 901.)

885^a. V$_2$. **Schwartz (A. W.)** proposes a blasting cartridge containing two vessels filled with compressed ammonia and chlorine gases respectively. By means of a mechanical arrangement or of a small charge of powder, two valves are opened simultaneously, allowing the two gases to mix. The object is to form chloride of nitrogen. (Spec. No. 7,098, 10.4.94.)

886. II. **Schwarz Powders** are blasting powders not readily ignited. Their composition is as follows :—

Saltpetre - - - - -	43 to 56 per cent.
Nitrate of soda - -	26 „ 18 „
Sulphur - - - - -	9 „ 10 „
Charcoal - - -	14 „ 15 „

(D. 607.)

887. III$_1$. **Sebastine** consists of:—

Nitro-glycerine	-	78 or 68 per cent.
Charcoal	-	14 ,, 20 ,,
Saltpetre	-	8 ,, 12 ,,

The charcoal is to be as porous and inflammable as possible. (D., p. 720, and Spec. No. 4,075, 21.10.76.)

888. **A.** III$_2$. **Securite,** like Bellite and Roburite, consist of a mixture of meta-dinitro-benzol with nitrate of ammonia, or nitrate of potash. Analysis of two samples gave the following proportions:—

I.		
Meta-dinitro-benzol	-	15·25 per cent.
Nitrate of ammonia	-	84·25 ,,
II.		
Meta-dinitro-benzol	-	25·18 ,,
Nitrate of potash	-	74·82 ,,

It is a yellow material with the odour of nitro-benzol. Other varieties contain trinitro-benzol or di or trinitro-naphthaline. The following equations are given as representing the various compositions and their products of combustion, and claim to show that it is a perfectly safe explosive to use in the presence of fire-damp and coal-dust:—

$$C_6H_4 . 2\,NO_2 + 10\,(NH_4 . NO_3) = 6\,CO_2 + 22\,H_2O + N_{22}$$
$$2\,(C_6H_3 . 3\,NO_2) + 15\,(NH_4 . NO_3) = 12\,CO_2 + 33\,H_2O + N_{36}$$
$$C_{10}H_6 . 2\,NO_2 + 19\,(NH_4 . NO_3) = 10\,CO_2 + 41\,H_2O + N_{40}$$
$$2\,(C_{10}H_5 . 3\,NO_2)\ \ 33\,(NH_4 . NO_3) = 20\,CO_2 + 71\,H_2O + N_{72}$$

(Paper by S. B. Coxon, at North of Eng. Inst., Mining and Mech. Eng., 11.2.87.)

A variety known as Flameless Securite (F. Schöneweg) consists of a mixture of nitrate and oxalate of ammonia and dinitro-benzol. It has been submitted for license and approved.

889. V$_1$. **Sederolit.** A mixture of chlorate of potash, sulphide of antimony, and sulphur. This is simply a detonating powder. (H. M.)

890. III$_1$. **Selenitic Powder** is a mixture of nitro-glycerine with plaster of Paris. (T., p. 100.)

891. **Selenophanite.** *See* PANCLASTITE.

892. III_2. **Selwig** and **Lange** have patented a method for the nitration of cotton, cellulose, straw, &c., consisting in carrying out the operation in a centrifugal machine, by means of which the acid is separated immediately the nitration is complete. (Fr. Spec. No. 213,983, 8.6.91.)

893. IV_1. **Seranine** is a mixture of nitro-glycerine and chlorate of potash. (D., p. 722.)

894. **Settle.** *See* WETTER.

895. **Shaem.** *See* SCHULTZE.

896. V_1. **Sharp** and **Smith's Powder** consists of :—

Saltpetre - - - - - - -	2 parts.
Chlorate of potash - - -	2 „
Yellow prussiate of potash - - -	1 „
Potash - - - - - - -	1 „
Sulphur - - - - -	2 „

(Spec. No. 2,779, 27.10.66. D. 613.)

897. **Short.** *See* KELLOW.

898. **Siegert.** *See* SILESITE.

899. IV_2. **Siemen's Powder** is a mixture of saltpetre, chlorate of potash, and a solid hydrocarbon, such as paraffin, pitch, caoutchouc, &c. The compound is mixed by sieves, and then treated with a volatile liquid hydrocarbon. A plastic mass is produced, which is made into cakes, rendered hard by evaporating the solvent, and granulated. It is stated to be non-explosive when unconfined. (Spec. No. 1,969, 26.4.82.)

900. **Sierch.** *See* KUBIN, WETTER.

901. **Siersch.** *See* ECRASITE.

902. III_2. **Siersch Powder,** an Austrian smokeless powder, analogous to Schwab Powder.

903. V_1. **Silesite** (Pietrowicz and Siegert) consists of :—

Chlorate of potash - - - -	60 per cent.
Sulphide of antimony - - - -	10 „
Sugar - - - - - -	30 „

Obviously a very sensitive compound.

904. R. III_2. **Silotvar** (W. von. Ruckterschell) is wood fibre treated with equal parts of nitric and sulphuric acids. It is therefore simply a weak nitro-cellulose. It looks like dry, flocculent sponge. It was submitted for licensing, but rejected as not standing the heat test. (Spec. No. 4,349 of 1886.)

905. IV_2. **Sjöberg** has patented :—

(1.) The manufacture of an explosive by dissolving nitrate of ammonium in a solid hydrocarbon, gelatinising in a liquid hydrocarbon, and adding chlorate of potash also gelatinised.

(2.) The application of gelatinisation to all ammonia salts, by dissolving in a solid hydrocarbon liquified.

(3.) The substitution for chlorate of potash of nitrate of ammonia, mixed with caseine, lactine, or other equivalent body. (Fr. Spec. No. 173,482, 12.1.86, 2.9.87.)

906. IV_2. **Sjöberg's Explosive** consists of :—

Nitrate or oxalate of ammonia	50	per cent.
Astral oil	10	,,
Naphthaline	5	,,
Chlorate of potash	35	,,

Sometimes five parts of chlorate of potash are replaced by an equal amount of carbonate of ammonia. (M. XX., p. 300.)

907.* III_2. **Sjöberg** has also, under the name of Nitro-lactose, patented a mixture of nitrated lactine and nitrated molasses, with the addition of nitrate of soda. (Fr. Spec. No. 192,683, 30.8.88.)

907*. **Sjöberg.** *See* NITRO-CURD.

908. **S.K. Powder.** *See* SMOKELESS POWDER (915).

909. III_2. **Skogland** has patented an explosive consisting of nitro-cellulose, or picric acid, with tartarate, carbonate, oxalate, or carbamate of ammonia, or other volatile base or hydroxyl. (Spec. No. 18,362, 15.12.88. Fr. Spec. No. 194,905, 20.12.88.)

910. **Skogland.** *See* GRAKRULT.

911. IV_2. **Sleeper's Powder** is a mixture of chlorate of potash, sugar, and charcoal.

912. **Smith.** *See* ELECTRIC FUZES.

913. **Smith.** *See* SHARP.

914. **Smokeless Explosive.** *See* ABEL (1).

915. **A.** III_2 **Smokeless Powder,** as manufactured by the Smokeless Powder Company, is defined in the same terms as Schultze Powder, viz.:—

"Consisting of nitro-lignin carefully purified and mixed or impregnated with a nitrate or nitrates (other than nitrates of lead or ammonium), and with or without starch, collodion, or turmeric, or other vegetable colouring matter, provided that such collodion shall consist of carefully purified nitro-lignin, dissolved in a safe and suitable solvent."

The powders manufactured are designated: S.R., S.S., S.K., S.V.

A sample of S.R. Powder gave:—

Nitro-cellulose (insoluble)	54·52	per cent.
„ „ (soluble)	19·92	„
Nitrates of baryta and potash	21·08	„
Matters soluble in ether	2·41	„
Moisture	2·07	„

See also RIFLEITE.

916. **A.** III_2. **Smokeless Blasting Powder** is defined as:—

Consisting of smokeless powder as above defined, with the addition of any one or more of the following ingredients, viz.:—Dinitro-benzol, dinitro-toluol, nitro-benzol, and nitro-naphthaline, provided that all such ingredients shall be thoroughly purified.

A sample gave:—

Nitro-lignin	23·84	per cent.
Nitro-compound	50·30	„
Nitrate of barium and potassium	25·48	„
Moisture	0·38	„

917. **Smokeless Powders.** *See* INTRODUCTION, p. xliii.

918. III $_{1}$., V_1. **Smolianoff** has patented a mixture of nitro-glycerine, and an alcohol or methyl-alcohol, with or without the addition of a fulminate. (Fr. Spec. No. 206,976, 15.7.90.)

919. **Smolianoff.** *See* AMERICANITE.

920. **A.** VII_2. **Snaps for Bonbon Crackers** are licensed as toy fireworks. They consist of a minute quantity of fulminate of silver, contained between two strips of card, to the ends of which friction patches are affixed.

921. III_1. **Snyder's Explosive** is an American dynamite, containing 6 per cent. of an absorbent, consisting of gun-cotton, collodion, and camphor dissolved in ether.

922. III_1. **Société Anonyme des Poudres et Dynamite** have patented explosives composed of nitro-cellulose, nitro-toluol, amylene, and nitro-glycerine, with or without the addition of nitrates or chlorates, and treated as follows:—

To one part of nitrate of cellulose add about its own weight of nitro-toluol (sp. gr. 1.16), so as to soften and partly dissolve the nitro-cellulose. In order to facilitate the process, 15 per cent. to 20 per cent. of amylene or potato oil is added. Nitrates or chlorates can be added. (Fr. Spec. No. 183,828, 26.5.87.)

923. III_1. **Société Français des Explosive** have proposed (in 1887) as an absorbent for nitro-glycerine a mixture of 50·81 parts of nitrated straw paper and 49·19 parts of nitro-toluol. (P. & S. 936.)

924. **A.** VII_2. **Socket Sound Signal** consists of a tin case, containing a charge of tonite, and having attached to its base a small charge of gunpowder. The signal is placed in a mortar and the gunpowder charge is fired. This ignites a time fuze which in its turn fires a detonator in the tonite charge.

925. **A.** VII_2. **Socket Distress Signal** is similar to the Sound Signal, but contains in addition coloured stars.

926. A. VII$_2$. **Socket Light Signal** consists of a tin cylinder containing white or coloured stars, and having attached to its base a small charge of gunpowder. It is fired in the same way as the Socket Sound Signals.

927. **Solvents.** The solvents most used in the manufacture of smokeless powders and other explosives are :—

Acetone - -	C_3H_6O.	
Acetic ether - -	$C_2H_5 . C_2H_3O_2$ (acetate of ethyl).	
Amylic ether - -	$C_5H_{11} . C_2H_3O_2$ (acetate of amyl).	
Alcohol - - -	C_2H_5OH	mixed together.
Ether - - -	$(C_2H_5)_2O$	

928. **Soulages.** *See* Safety Blasting Powder.

929. A. VII$_2$. **Sound Signal Rocket** consists of a signal rocket with one or more charges of tonite, containing detonators, fitted in the head.

930. I. **S.P. Powder** (Selected Pebble) the designation of Lots of pebble powder selected from the results of proof, for certain guns. S.P.$_1$, S.P.$_2$, and S.P.$_3$ are the designations of powders in the French Service.

931. **Spence.** *See* Ricker.

932. **Spon.** *See* Electric Fuzes.

933. **Spooner.** *See* Nitro-flax.

934. **Sprengel.** *See* p. xli.

935. III$_2$. **Springthorpe** has proposed as a smokeless powder an explosive which is simply nitrated oat straw or wood fibre. (Fr. Spec. No. 207,919, 29.8.90)

936. **S.R. Powder.** *See* Smokeless Powder (915).

937. **S.S. Powder.** *See* Smokeless Powder (915).

938. **S.T. Dynamite** is identical with E.C. Dynamite.

938*. A. III $_1$. **Standard Dynamite, No. 1** consists of:—

Nitro-glycerine - - -	60 per cent.
Nitro-cotton - - - -	40 ,,

No. 2 variety contains nitrates in place of a portion of the nitro-cotton.

939. II. **Starch Powder.** To the ordinary ingredients of gunpowder 2 to 5 per cent. of starch is added. If the charcoal and sulphur be boiled in a solution of nitre and starch, a hard and less hygroscopic powder is produced; but, like all powders prepared in this way, the mixture is not uniform. (O. G.)

940. II. **Stones,** like Oliver, whom he considerably preceded, also proposed to use charred peat in the manufacture of gunpowder. (Spec. No. 12,990, 7.3.50.)

941. A. III $_1$. **Stonite,** is defined as:—

"Consisting of not more than 68 parts by weight of thoroughly purified nitro-glycerine, uniformly mixed with 32 parts by weight of a preparation consisting of nitrate of barium, nitrate of potassium (or either of them), kieselguhr (not less than 20 parts by weight), wood meal (not less than four parts by weight), and carbonate of magnesia, with or without the addition of sulphuretted oil and soot (or either of them), such preparation to be sufficiently absorbent when mixed in the above proportions to prevent the exudation of nitro-glycerine."

942. II., IV $_2$. **Storer** proposes the use of kerosene shale mixed with an oxidising agent. (Spec. No. 23,377, 28.1.93.)

943. IV $_2$. **Storite,** as submitted for license in the Colony of Victoria and rejected, consists of:—

Chlorate of potash - - -	70 per cent.
Brown coal shale - - -	30 ,,

This is clearly the same explosive as the preceding one.

943*. III $_1$. **Straw Dynamite** is a mixture of nitro-glycerine with nitro-cellulose made from straw. It is termed Paleine. *See also* LANFREY'S POWDER. (Spec. No. 3,119, 7.8.78.)

944. III $_2$. **Studer Powder,** a nitro-powder for use in small bore rifles. It is probably composed mainly of nitro-cellulose.

945. **Sulphide of Nitrogen.** *See* NITROGEN SULPHIDE.

946. **Sulphurite.** *See* SALA and AZEMAR.

947. **S.V. Powder.** *See* SMOKELESS POWDER (915).

948. III$_1$. **Terrorite,** proposed by Mindcleff for use in shells, consists of nitro-glycerine and methyl-alcohol in varying proportions. It was tried by the Mexican Government, but the results were unsatisfactory.

948*. **Tetranitro-.** *See* NITRO-.

949. **Teutonite.** *See* AUGENDRE.

950. III$_1$. **Thorn (L. T. G.)** has proposed a smokeless powder, consisting of about 40 parts of nitro-cresol with 20 parts of carbonate or nitrate of baryta or strontia. The whole is treated with a solution of liquid resin, wax, &c., in alcohol, so as form a plastic mass, which is dried and converted into grains. (Spec. No. 16,189, 11.10.90. Fr. Spec. No. 208,796, 11.10.90, 18.2.91.)

951. III. **Thorn, Westendarp** and **Pieper** have patented smokeless explosives consisting of nitro-cresols or their salts, with or without nitrates of baryta and strontia. (Fr. Spec. No. 208,796, 11.10.90, 18.2.91.)

952. **A.** VII$_2$. **Throwdowns,** a toy firework consisting of a patch of fulminate of silver, wrapped in a piece of tissue paper containing fragments of flint.

953. IV$_1$. **Thunder Powder** consists of honey and glycerine treated with nitric and sulphuric acids. The product is mixed with chlorate of potash, nitrate of potash, prepared sawdust, and prepared chalk in two proportions for two different grades. (T., p. 107.)

954. III$_1$. **Thunderbolt Powder,** an American form of dynamite of the No. 2 type. (T. 88.)

955. III$_1$. **Titan Powder,** an American form of dynamite of the No. 2 type. (T. 88.)

956. III$_2$. **Titan Powder** consists of vegetable fibre, pulped, compressed, granulated, and treated with the usual acids. Also vegetable fibre prepared with a solution of sugar, or mannite, or amyline,* or inuline,† and then treated with acids. This last gives obviously a mixture of various nitro-compounds. (T., p. 102.)

957. **A.** III$_2$. **Tonite** or **Cotton Powder** is a nitrated gun-cotton, the nitrate usually employed being that of barium. It is usually issued in the form of cartridges, coated with paraffin of a brown hue. The No. 2 variety is the same as the above, with the addition of charcoal, which gives the resulting compound a grey appearance.

A sample of Tonite No. 1 gave :—

Gun-cotton - - - - -	50·90 per cent.
Nitrate of baryta - -	48·71 „
Moisture - - - - - -	0·39 „

See POTENTITE.

958. **A.** III$_2$. **Tonite No. 3** consists of gun-cotton, dinitrobenzol, chalk, and nitrate of potash, soda, or baryta.

A sample gave :—

Gun-cotton - - - - -	14·55 per cent.
Meta-dinitrobenzol - -	13·20 „
Nitrate of baryta - - - -	72·25 „

959. III$_2$. **Tonkin** proposed to mix together :—

Pulped gun-cotton - - - - -	3 per cent.
Nitrate of potash or soda - -	56 „
Charcoal - - - -	26 „
Sulphur - - - - - - -	15 „

In some cases the cotton was to be employed in an unconverted state. The compound was a sort of gunpowder, with 3 per cent. of gun-cotton. It was to be granulated.

960. **Toy Caps.** *See* AMORCES.

---

* Amyline (C_5H_{10}) is a transparent, colourless, very thin liquid, produced by the dehydration of amylic alcohol ($C_5H_{12}O$) or fusel oil.

† Inuline ($C_6H_{10}O_5$) is a substance much like starch, prepared from various plants.

961. III$_1$. **Trauzl's Dynamite** is a mixture of nitro-glycerine and gun-cotton pulp. One sample consisted of:—

Nitro-glycerine - - - -	75 parts.
Gun-cotton - - - - -	25 „
Charcoal - - - - - - -	2 „

This compound is stated to have been detonated by a strong detonator after being four days under water and containing 15 per cent. of moisture. (D., p. 725.)

962. **Travers.** *See* ACETYLIDES.

963. **Trench.** *See* MACKIE, OARITE, ROBURITE.

964. III$_2$. **Trench** has also proposed for fiery mines to surround tonite cartridges with a mixture of sal ammoniac, common salt, alum, and wood meal, the whole being enveloped in a paper sheath. *See also* WETTER.

965. **Tribenite.** *See* CADORET.

966. **Trinitro-cellulose, -cresol, -phenol,** &c. *See* NITRO-CELLULOSE, &c.

967. III$_2$. **Trinitro-toluol** has been found by Häusserman to be a powerful explosive when fired with a detonator, but difficult to explode in any other way. (Journal of Society of Chemical Industry, X. 1,028.) *See also* NITRO-TOLUOL.

968. **Triumph Safety Powder.** *See* COURTEILLE'S POWDER.

969. **A.** III$_2$. **Troisdorf Powder** is defined as consisting of gelatinised nitro-cellulose with or without the addition of nitrates.

970. III$_2$. **Trotman** proposed to mix gun-cotton or other nitro-cellulose, in the condition of pulp or powder, with silicate cotton.* The object is to retard the explosion of gun-cotton by mixing with it an inert, non-hygroscopic material. The proportions recommended are 25 parts of silicate cotton to 75 parts of gun-cotton. (Spec. No. 2,536, 24.6.79.)

971. **Trützschler-Falkenstein.** *See* HIMLY.

---

* Produced from blast furnace slag by blowing into it when melted. It is much used as a non-conducting material for packing steam pipes, &c.

972. IV_2. **Tschirner's Powder** consists of 57 parts of picric acid and 43 parts of chlorate of potash incorporated together with 5 per cent. of powdered resin. The product is sprinkled with benzol or kerosine to moisten it and dissolve the resin. The compound becomes a plastic mass, easily moulded, and the solvent evaporates off. (Spec. No. 447, 31.1.80, and No. 3,846, 22.9.80.)

973. A. VI_1. **Tube Safety Fuze** is a variety of metallic safety fuze.

974. A. VI_3. **Tubes for Firing Explosives** are cases of quill, metal, or paper charged with mealed powder or other suitable explosive. They are classed under Division 3 or Division 2 of Class VI., according to whether they contain their own means of ignition or not.

975. R. IV_2. **Turpin's Powders** consist of mixtures of chlorate of potash with coal tar and charcoal. In one variety about half the chlorate is replaced by nitrate of potash. Some absorbent of the nature of charcoal, silica, kieselguhr, &c. is added according to the degree of fluidity of the coal tar. The sensitiveness of the compound is increased by substituting 1 to 10 per cent. of permanganate of potash for an equivalent amount of the chlorate. The following proportions are recommended:—

—	No. 1.	No. 2.
Chlorate of potash - - -	80 per cent.	40 per cent.
Coal tar - - -	14 to 16 ,,	15 ,,
Wood charcoal - - -	6 ,, 4 ,,	5 ,,
Nitrate of potash - -	— ,,	40 ,,

Nitrate of lead may be substituted for saltpetre. (Spec. Nos. 4,544, 18.10.81, and 2,139, 27.4.83.)

976. III_2. **Turpin** has patented the application of the explosive properties of picric acid to industrial and military uses. (Fr. Spec. No. 167,512, 7.2.85.) In two "certificates of addition" (17.10.85 and 1.9.92) he claims methods of charging and detonating shells filled with picric acid.

977. III_2. **Turpin** has also patented the use of the explosive properties of chloro-bromo-iodo-nitro bodies without oxydising agents, by detonation. (Fr. Spec. No. 185,029, 27.7.87.)

978. III_2. **Turpin** has also patented the use of organic nitro-amide bodies in the manufacture of explosives, as follows:—

Picric or cresylic acid is boiled with iron or zinc scrap, and with the addition of acetic acid or by other means, until picramic acid is formed. This, either alone or in combination with a base (picramate of soda or potash), constitutes an explosive.

He suggests the following compositions:—

1.

Picramic acid	30 per cent. to	50 per cent.
Nitrate of potash	70 „ „	50 „

2.

Picramate of soda	25 per cent. to	53 per cent.
Nitrate of potash	75 „ „	47 „

(Fr. Spec. No. 185,034, 27.7.87.)

979. III_2. **Turpin** has also patented, under the name of "Poudre Celluloïque" or "Celluloïdine," an explosive prepared by dissolving gun-cotton in a volatile solvent, and spreading the paste in sheets, which are cut into grains after the solvent has evaporated. (Fr. Spec. No. 189,398, 16.3.88.)

980. $III._{1 \& 2}$., IV_2. **Turpin** has also patented the following mixtures which can be used without the aid of a detonator:—

(1.) *Progressite.*

Nitrate of baryta	65 per cent.
Picrate of ammonia	15 „
Dinitro-benzol	10 „
Coal tar	6 „
Brown charcoal	4 „

(2.) *Duplexite.*

Chlorate of potash or baryta	70 „
Charcoal	10 „
Dinitro-benzol	10 „
Coal tar	10 „

Turpin—*continued.*

(3.)

Nitrate of baryta - - -	60 per cent.
Dinitro-benzol - - -	15 ,,
Nitro-phenol - -	25 ,,

(4.)

Nitrate of baryta - - -	60 ,,
Picramate of soda - - -	20 ,,
Nitro-benzol - - - -	10 ,,
Nitro-phenol - - -	10 ,,

(Fr. Spec. No. 189,426, 17.3.88.)

981. III $_{1.\ \&\ 2.}$, IV $_2$. **Turpin** has also patented the following mixtures as flameless explosives :—

(1.) *Boritine.*

Nitro-glycerine - - - -	7·5 parts.
Kieselguhr, &c. - - -	2·5 ,,
Boric acid - - - -	10·0 ,,

(2.) *Fluorine.*

Nitro-glycerine - - -	7·5 ,,
Some absorbent - - -	2·5 ,,
Fluoride of calcium - -	10·0 ,,

(3.) *Oxydine.*

Nitro-glycerine - - - -	7·5 ,,
Some absorbent - - -	2·5 ,,
Oxide or sulphide of zinc - -	10·0 ,,

(4.)

Chlorate of potash - - -	70·0 ,,
Charcoal - - - -	10·0 ,,
Dinitro-benzol - - -	10·0 ,,
Coal tar - - - -	10·0 ,,
Boric acid - - - -	100·0 ,,

(5.)

Dinitro-benzol - - -	30 ,,
Some nitrate or chlorate - - -	70 ,,
Fluoride, &c. - - -	100 ,,

(Fr. Spec. No. 189,428, 17.3.88.)

982. VI $_2$. **Turpin** has also patented a method of charging shells with picric acid and similar bodies, without the aid of any oxidizing agent. (Fr. Spec. No. 205,429, 3.5.90.)

983. **Turpin.** *See* MELINITE, PANCLASTITE, and p. xxxix.

983[a]. **Uchatius.** *See* NITRO-STARCH.

984. $III_{1\cdot\ \&\ 2\cdot}$ **United States Smokeless Powder Company** have patented an explosive, consisting of picrate of ammonia and nitrate of ammonia in varying proportions, with or without the addition of nitro-glycerine. (Spec. No. 12,415, 22.10.92.)

985. III_{2}. **United States Smokeless Powder Company** have also patented the following:—

Bichromate of ammonia - -	20 per cent.
Picrate of potash - - -	25 „
Picrate of ammonia - - -	55 „

(Spec. No. 1,982, 3.3.94.)

986. **Vegetable Powder.** *See* CASTAN.

987. III_{1}. **Vending** in 1882 patented an explosive which he called "Dynamite Nitro-benzoique," composed as follows:—

Nitro-glycerine - - -	15 to 45 parts.
Nitro-cellulose - - - - -	1 „ 3 „
Nitro-benzol - - -	5 „ 10 „
Nitrate of ammonia - - -	50 „ 73 „

988. **R.** III_{1}., IV_{2}. **Victorite** proposed by Punshon somewhat resembles Tschirner's Powder. It consists of chlorate of potash, picric acid, and a little olive or other oil, with the occasional addition of some charcoal. It is in the form of a coarse yellowish grey powder, which leaves an oily stain on paper. It is extremely sensitive to friction and percussion. It was submitted for approval in this country and rejected.

In the specification the following proportions are given:—

Chlorate of potash - - -	80 parts.
Picric acid - - - - -	110 „
Nitrate of potash, soda, or baryta -	10 „
Charcoal - - - - -	5 „

Another variety substitutes nitro-glycerine for chlorate of potash. (Spec. No. 11,140, 1887.)

989. III_{2}. **Vieille Powder,** a smokeless powder adopted by the French Government. It is in the form of thin horny tablets, which have been probably prepared by dissolving gun-cotton in a volatile solvent. In the earlier samples it is believed that picric acid was added as an ingredient. This combination, however, is very unstable.

990. **Vieillard.** *See* MAGNIER.

991. **Vigorite.** *See* BJORKMANN (E. A.).

992. III$_{1}$., IV$_{1}$. **Vigorite** consists of about:—

Nitro-glycerine - - -	30 per cent.
Nitrate of soda - - - - -	60 ,,
Charcoal - - - -	5 ,,
Sawdust (or partly nitrated wood or paper pulp) - - - -	5 ,,

Another brand as manufactured by the California Vigorite Powder Company consists of:—

Nitro-glycerine - - -	43·75 per cent.
Nitrate of potash - - - -	18·75 ,,
Chalk - - - -	8·75 ,,
Sawdust - - -	11·25 ,,
Chlorate of potash - -	17·50 ,,

The presence of the deliquescent nitrate of soda renders this a specially dangerous mixture, for the salt melts readily when exposed to damp, and the nitro-glycerine consequently exudes. A serious accident was occasioned by vigorite made by the Hamilton Powder Company on the 5th May 1879, when a consignment of it, described as blasting powder, exploded on the Grand Trunk Railroad at Stratford, Ontario, while the cars were being shunted. Two men were killed and several injured, while 24 cars were destroyed and 100 damaged.

993. **Viner.** *See* WIENER.

994. II. **Violette's Powder** consists of:—

Nitrate of soda - - -	62·5 per cent.
Acetate ,, - - - - -	37·5 ,,

The ingredients can be melted together so as to give a close combination, but if heated to about 660° F. the mixture explodes. It is hygroscopic. (B. II., p. 315.)

995. **A.** III$_{1}$. **Virite No. 1** consists of:—

Nitro-glycerine - - -	5 parts.
Charcoal - - -	9½ ,,
Nitrate of potash - - -	33½ ,,

995ª. **R.** III$_{1}$. **Virite No. 2** contained nitrate of sodium in place of nitrate of potash, and as might have been expected (*see* VIGORITE) failed to retain its nitro-glycerine under conditions which might well arise during ordinary transport, and it was therefore rejected.

996. **Vizer.** *See* PUNSHON and VIZER.

997. **Vogt.** *See* GIRARD.

998. II. **Volkman** has proposed a powder consisting of a mixture of saltpetre, wood meal, and ferrocyanide of potash. He has named the variety for blasting "nitropyline" and that for sporting purposes "collodine." (P. & S. 1,012.)

999. IV_1. **Volney's Explosive** is produced by saturating concentrated glycerine ($C_3H_5(OH)_3$) with hydrochloric acid gas to form glycerine chlorhydrin ($C_3H_5 . (OH)_2Cl_2$) which is treated with a mixture of nitric and sulphuric acids to form mono-chlordinitrin (C_3H_5 Cl . $2NO_2$). Chlorates are added. (T., p. 107.)

1000. III_2. **Volney's Powders** consist of :—

No. 1.

Di-and trinitro-naphthaline	218 parts.
Saltpetre	19 „
Sulphur	16 „

No. 2.

Mono-nitro-naphthaline	100 parts.
Saltpetre	330 „
Sulphur	51· „

Any nitrate or chlorate can be substituted for saltpetre. (M. XX. 298.)

1001. IV_2. **Von Brank's Powders** consist of :—

Shooting Powder.

Chlorate of potash	100 parts.
Boiled linseed oil	40 „
Lead oxide	1·5 „

Blasting Powder.

Chlorate of potash	100 parts.
American resin	15 „

(Spec. No. 5,027, 20.3.91.)

1002. III_2. **Von Brauk** proposed a smokeless powder consisting of gun-cotton and carnauba wax mixed together when warm.

1003. **Von Förster.** *See* WOLFF and VON FÖRSTER.

21

1004. A. III_2. **Von Förster's Smokeless Powder** is defined as consisting of nitro-cellulose gelatinised by a suitable process, with or without the addition of carbonate of calcium and graphite.

1005. III_2. **Von Freeden** has patented a method of granulating mixtures containing nitro-cellulose by agitating the gelatinised mass in presence of a liquid or vapour not exercising any chemical action upon it. (Fr. Spec. No. 203,734, 12.2.90.)

1006. **Von Ruckterschell.** *See* SILOTVAR.

1007. III_1. **Vonges.** *See* DYNAMITE DE VONGES.

1008. R. IV_2. **Vril** consists of (approximately):—

—	1.	2.
	Per Cent.	Per Cent.
Chlorate of potash	50·0	48·0
Yellow prussiate of potash	4·5	9·1
Nitrate of potash	25·0	24·3
Willow charcoal	12·5	11·6
Paraffin	6·0	6·5
Ferric oxide	2·0	0·5

(No. 2 variety is less sensitive than No. 1.)

In external appearance it resembles ordinary gunpowder. It was proposed to be licensed as an authorised explosive, but failed to satisfy the requirements of safety.

1009. R. III_1. **Vulcan Dynamite** is simply a lignin-dynamite containing nitrate of soda. (P. & S. 1020.)

1010. III_1. **Vulcan Powder** is of the following approximate composition:—

Nitro-glycerine	30	per cent.
Nitrate of soda	52·5	,,
Sulphur	7	,,
Charcoal	10·5	,,

It is much the same as Vigorite or Virite No. 2. It was largely used in the first Hellgate explosion at New York, 11,852 lbs. of it being used in conjunction with 9,127 lbs. of rendrock and 28,935 lbs. of dynamite. (D., p. 721, and "Abel on Explos. Agents," p. 26.)

1011. **Vulcanienne.** *See* ESPIR.

1012. V_{1}., III_{2}. **Vulcanite** (Moritz and Köppel) is composed as follows:—

—	No. 1.	No. 2.
Nitrate of potash - - -	35 parts.	42 parts.
Nitrate of soda - - - -	19 "	22 "
Sulphur - - - -	11 "	12·5 "
Wood meal - - - - -	9·5 "	10 "
Chlorate of potash - - -	9·5 "	—
Charcoal - - - - - -	6 "	7 "
Sulphate of soda - - - -	2·25 "	5 "
Sugar - - - - - -	2·25 "	—
Picric acid - - - -	1·25 "	1·5 "

It is an Austrian explosive. (O. G.)

1013. **Waffen.** *See* LEDERITE.

1014. III_{1}. **Waffen** has proposed the following:—

Nitro-glycerine - - -	37·6	per cent.
Collodion cotton - - -	2·4	"
Nitrate of soda - - -	22·5	"
Wood meal - - - -	36·0	"
Picric acid - - -	0·25	"
Sulphur - - - - -	1·0	"
Carbonate of soda - -	0·25	"

(M. XXX. 503.)

1015. $IV_{1\ \&\ 2}$. **Wahlenberg** proposed an explosive consisting of mono-, di-, or trinitro-benzol, chlorate of potash, and alkaline nitrates, the latter being preferably treated with solid hydrocarbons to make them non-deliquescent. Nitrate of ammonia was specially recommended, and considerable safety is claimed for the compound. (Spec. No. 2422, 12.6.76.)

1016. **Walker.** *See* GALLAHER.

1017. A. III_{2}. **Walsrode Powder** is defined as consisting of thoroughly purified nitro-cellulose, mixed with carbonate of calcium, and gelatinised by a suitable process.

1018. **Walthamite.** *See* CANNONITE.

1019. $III_{1\,\&\,2}$. **Wanklyn** has patented an explosive consisting of one part of nitrate of urea, and two to five parts of gun-cotton, dynamite, or other nitro-compound. (Spec. No. 9,799, 5.7.88. Fr. Spec. No. 199,375; 4.7.89.)

1020. **A.** VI_2. **War Rockets** are large rockets with heavy metal heads. They are licensed only when they do not contain their own means of ignition.

1021. **Ward.** *See* GRAHAM.

1022. V_1. **Ward** and **Gregory's Priming Composition** consists of :—

Chlorate of potash - - -	75 parts.
Amorphous phosphorus - - -	1 ,,
Coke - - - - -	1 ,,

The ingredients are mixed together with the aid of a volatile liquid. Sometimes paraffin or tallow oil are added. (A. & E. I. 110.)

1023. III_1. **Warren's Powder** consists of one part of nitro-cellulose added to 10 parts of nitro-glycerine. The mixture is allowed to stand until the nitro-cellulose is dissolved without heat. Pulverised trinitro-cellulose is mixed in till the mass is brought to the consistency of a dry powder. Add pressed and glazed gunpowder, the proportion varying with the strength required. A proportion of 70 parts gunpowder to 30 parts of the above mixture is suggested. "One of the main objects of this "compound is to keep the added gunpowder dry." (T., p. 106.)

1024. III_2. **Wass Powder** consists of nitro-cellulose partially gelatinised and mixed with sesqui-oxide of manganese.

1025. **Wasserfuhr.** *See* COLOGNE POWDER.

1026. **Watson.** *See* DAVEY.

1027. **W.** *See* HALL.

1028. **Wellite.** *See* HEBLER.

1029. III_2. **Wellorech's Powder** is described as a "pure nitro-cellulose powder." (A. & E. II. 84.)

1030. **Weniger.** *See* PREISENHAMMER.

1031. A. II. **Westfalite** is a dark yellowish-buff powder consisting of nitrate of ammonia and gum-lac soluble in alcohol. It is manufactured by grinding the nitrate of ammonia in an alcoholic or other solution of the gum-lac, and then evaporating the solvent. The proportions recommended are:—

Ammonium nitrate - - -	90 parts.
Gum-lac - - - - - - -	6½ „

(Spec. No. 15,566, 23.9.93.)

1032. III_1. **Wetter Dynamite** (Müller and Aufschläger). The principle of this is the incorporation of salts having a high proportion of water of crystallisation with dynamite, so as to be safe for use in fiery mines. (Spec. No. 12,424, 13.9.87.)

A mixture of 40 per cent. of soda crystals with dynamite or of alum with gelatine dynamite may be used. Soda with gelatine dynamite gives a substance as hard as stone, and cannot therefore be used.

The French Committee at Sevran-Livry, in 1888, tried various mixtures. Amongst them were equal weights of dynamite and soda crystals, or sulphate of soda, or ammonia alum, or ammonium chloride. They stated that a temperature of 2,000° C. was required to explode fire-damp, and gave the explosion temperatures of dynamite, nitro-glycerine, and gun-cotton, at 2,940°, 3,170°, and 2,636° respectively. Similar mixtures are 20 per cent. of dynamite, &c., to 80 per cent. of nitrate of ammonium. Estimated temperature of explosion 1,130° C. (Mallard and le Chatelier). Also Kubin and Sierch's mixture of dynamite, with 20 to 50 per cent. of ammonium chloride or sulphate, or both (Spec. No. 3,759, 10.3.88), and for a similar purpose Kuhnt and Diessler use 60 per cent. of carbonate of ammonia. (Spec. No. 5,949, 21.4.88.)

For use with tonite cartidges, Trench recommends an envelope of sal-ammoniac, salt, alum, and sawdust, saturated with water. The envelope is a paper case

Wetter Dynamite—*continued.*

completely surrounding the cartridge, and containing the mixture as above. Settle's water cartridge, Heath and Frost's gelatinous cartridge, and those of other inventors, are mechanical means directed to the same end.

The Anzin Company are stated to have consumed more than 30,000 kilos of "safety explosives," consisting of:—

—	A.	B.	C.
	Per Cent.	Per Cent.	Per Cent.
Gelatinised nitro-glycerine - -	20	30	—
Ammonium nitrate - - -	80	70	90
Octometric gun-cotton - -	—	—	10

and find the results favourable. ("Colliery Guardian," 30.5.90.)

1033. III $_2$. **Wetteren Powder,** a smokeless powder manufactured by Messrs. Cooppal & Co. It is in the form of thin horny tablets, and is composed almost entirely of gelatinised gun-cotton with the addition of a small per-centage of carbonate of lime.

1034. II. **Wetzlar Powder** is composed as follows:—

Nitrate of soda - - -	66·68	parts.
Sulphur - - - -	11·77	„
Spent tan - - - -	18·71	„

(D. 609.)

1035. **White Gunpowder.** *See* AUGENDRE, POHL, and REVELEY.

1036. I. **Wiener** proposed to mix the ingredients of gunpowder in a dry state, and subject them to action of a steam-heated press at a temperature of about 250° F. The sulphur melted at this temperature, and became thoroughly distributed throughout the mixture, converting the whole into a compact homogeneous black cake. (Spec. No. 3,731, 17.11.73.)

Wiener—*continued.*

Experiments were tried with this method of manufacture at Woolwich in 1878. The prospective advantages claimed were—

1. Decrease of the power of absorbing moisture.
2. Economy of manufacture, by doing away with stoves and mills.
3. Diminution of the danger of manufacture due to the presence of small quantities of gunpowder in any one working building.

The English experiments failed, however, to produce a satisfactory powder, and it was ascertained that in the manipulation by Wiener (Colonel Viner) the mixture had been raised to the fusing point of saltpetre, which approximates to the firing point of gunpowder, and that several explosions had occurred. Consequently the matter was dropped, but a few experiments were tried with ordinary powder heated to the melting point of sulphur, the idea being to coat each grain with a film of sulphur, or allow the latter to penetrate the charcoal. The trials were fairly successful, but not to such an extent as to warrant further experiments on a large scale. Gunpowder treated in this way was called "Baked Powder." (Reports of Com. on Explos. Sub., 1880–1.)

1037. V_1. **Wigfall's Powder,** or **Prussian Fire,** consists of (subject to variation):—

Carbon	4 parts.
Gums	4 ,,
Aquafortis	6 ,,
Red lead	40 ,,
Cannel coal	1 ,,
Steel filings	4 ,,
Phosphorus	4 ,,
Sulphur	2 ,,
Chlorate of potash	26 ,,
Sugar of lead	6 ,,
Saltpetre	3 ,,

It is doubtful from its nature whether the above very heterogeneous and somewhat alarming compound was ever actually made. (Spec. No. 2,888, 18.11.63.)

1038. **Willard.** *See* HERCULES.

1039. IV_2. **William's Powder** consists of:—

Chlorate of potash	48 parts.
Prussiate of potash	16 „
Bichromate of potash	2 „
Nut galls	5 „
Cannel coal	2 „
Starch	6 „
Crude coal oil	5 „

1040. II. **Windsor's Powder** consists of 25 parts of dry powdered sugar added to 100 parts of gunpowder. (Spec. No. 3,510, 4.12.11.)

1041. V_1. **Winiwarter's** fulminating compositions consist of:—

No. 1.

Fulminate of mercury	300 parts.
Chlorate of potash	288 „
Sulphide of antimony	312 „
Charcoal 16·7 } Saltpetre 65·3 }	60 „
Ferro-cyanide of potassium	23 „
Binoxide of lead	6 „
Solution of 75 parts pyroxilin in 150 parts of ether, called by him ether-oxylin	900 „

No. 2.

Fulminate of zinc	75 parts.
Chlorate of potash	12 „
Sulphide of antimony	7 „
Binoxide of lead	15 „
Ether-oxylin	224 „
Ferro-cyanide of potassium	1 „

No. 3.

Amorphous phosphorus	75 parts.
Binoxide of lead	64 „
Charcoal and saltpetre	15 „
Ether-oxylin	106 „

(Spec. No. 13,935, 29.1.52, and No. 306, 4.2.53.)

1042. III_1. **Wohanka** adds cellulose to the liquid explosives, made by dissolving in concentrated nitric acid the nitro-derivatives of the hydrocarbons of the aromatic phenol series. The cellulose becomes nitrated and swells up, forming with the explosive a plastic mass like gelatine. (Spec. No. 7,608, 25.5.87.)

1043. III $_2$. **Wolff** and **Von Förster** have patented a process consisting of cutting cubes or grains from slabs of highly compressed gun-cotton. (Fr. Spec. No. 164,792, 14.10.84.)

1044. A. III $_2$. **Wood Gunpowder,** a generic name for Nitro-lignin preparations. *See* also PATENT GUNPOWDER.

1045. **Woodnite.** *See* CHABERT.

1046. II. **Wynant's Powder.** In this powder nitrate of baryta wholly or partially replaces the saltpetre in gunpowder. The following proportions are recommended:—

Nitrate of baryta	- - -	77 per cent.
Charcoal	- - -	21 ,,
Saltpetre	- - - -	2 ,,

(This is identical with Saxifragine.)

The nitrates of lead or strontia may replace the nitrate of baryta. The inflammability of the powder may be increased by dusting or coating the grains with ordinary fine powder. (Spec. No. 1,084, 15.4.62.)

Experiments were made at Brussels with a powder in which $\frac{4}{5}$ of the saltpetre in gunpowder were replaced by nitrate of baryta. It was found to be unfit for small-arms on account of its slowness of combustion, and for larger arms on account of its fouling. It was consequently relegated to the rank of a blasting powder only. (D., p. 610.)

1047. II. **Xanthine Powder,** invented by Professor Schwar of Gratz, consists of a mixture in which the sulphur and charcoal of ordinary powder are replaced by a compound containing them both, *e.g.*, xanthate of potash (C_2 H_5 KCO S_2), a compound prepared by adding to absolute alcohol an excess of pure caustic potash and of bisulphide of carbon. The proportion is:—

Saltpetre	- - -	100 parts.
Xanthate of potash	- - -	40 ,,
Charcoal	- - -	6 ,,

(B. II., p. 315.)

1048. A. III $_2$. **Xylobrome** consists of nitro-lignin mixed with nitrates.

1049. III $_1$. **Xyloglodine** consists of glycerine and starch, or glycerine and cellulose, or glycerine and mannite, or glycerine and benzol, or analogous substances treated with the usual acids. It is claimed that it differs in certain characteristics from nitro-glycerine. (T., p. 101.)

1050. **Xyloidine.** *See* NITRO-STARCH.

1051. **Yates.** *See* HARRISON.

1052. **Yellow Powder.** *See* DARAPSKI.

1053. V $_1$. **Zaliwsky** proposed to mix chlorate of potash with oxalic acid before adding it to sulphur, charcoal, or other ingredients. The object was safety in manufacture and manipulation. (D., p. 616.)

1054. **Zanky Dynamite.** *See* KRUMMEL.

1055. **Zini.** *See* MAÏZITE.

1056. **Zschokke.** *See* COLLIERY SAFETY LIGHTERS.

INDEX TO INGREDIENTS.

www.ingramcontent.com/pod-product-compliance
Lightning Source LLC
Chambersburg PA
CBHW030318270326
41926CB00010B/1416